In Vitro

LABORATORY TECHNIQUES IN BIOCHEMISTRY AND MOLECULAR BIOLOGY

Edited by

P.C. van der Vliet — *Department for Physiological Chemistry, University of Utrecht, Utrecht, Netherlands*

Volume 26

ELSEVIER

AMSTERDAM – LAUSANNE – NEW YORK – OXFORD – SHANNON – SINGAPORE – TOKYO

ANALYSIS OF RNA-PROTEIN COMPLEXES *IN VITRO*

Jørgen Kjems
Jan Egebjerg

Department of Molecular and Structural Biology
Aarhus University
C.F. Møllers Allé, building 130
DK-8000 Århus C, Denmark

and

Jan Christiansen

Institute of Molecular Biology
University of Copenhagen
Sølvgade 83H
DK-1307 Copenhagen K, Denmark

1998
ELSEVIER
AMSTERDAM – LAUSANNE – NEW YORK – OXFORD – SHANNON – SINGAPORE – TOKYO

ELSEVIER SCIENCE B.V.
Sara Burgerhartstraat 25
P.O. Box 211, 1000 AE Amsterdam, The Netherlands

Library of Congress Cataloging in Publication Data
A catalog record from the library of Congress has been applied for.

ISBN: 0-444-82418-9 (pocket edition)
ISBN: 0-444-82419-7 (library edition)
ISBN: 0-7204-4200-1 (series)

Transferred to digital printing 2006

Printed and bound by CPI Antony Rowe, Eastbourne

Preface

The central role of RNA in many cellular processes, biotechnology, and as pharmaceutical agents, has created an increasing interest in experimental methods applied to RNA molecules. Two decades ago, RNA was generally considered a rigid molecule providing a framework for catalytically active proteins, or a passive messenger in the flow of genetic information. Today, we know that RNA is a highly active component, crucial in many catalytic and dynamic processes and a main target for post-transcriptional regulation of gene expression. Moreover, advances in biotechnology have enabled us to study many of the cellular processes *in vitro*.

This book covers most of the commonly used biochemical methods for investigating RNA-protein complexes *in vitro*. The protocols can be followed by most researchers trained in standard molecular biological techniques and require a minimum of specialised equipment. The methods are applicable to all classes of RNAs, and generally do not depend on whether the RNA has been prepared from natural sources or *in vitro*. However, due to variations in the behaviour of individual RNAs, optimisation may be required to improve the result of some of the methods. The discussion of critical parameters and possible pitfalls in the procedures provides a guideline for optimisation. Moreover, key references, where the reader can seek more specialised information, are listed at the end of each chapter.

This book is divided into five chapters. Chapter 1 is an introduction to the characteristics of RNA-protein interactions based on models that have emerged from the plethora of technologies available today. Chapter 2 describes the specific precautions recommended for working with RNA and outlines the most commonly used protocols for preparation of RNA from cells, or by *in vitro* techniques. Chapter 3 provides basic protocols for preparation of cellular extracts and recombinant bacterially expressed protein aimed at RNA-protein complex formation, and Chapter 4 focuses on the preparation and characterisation of the RNA-protein complex. Since many scientists are particularly interested in eukaryotic RNA metabolism, procedures for purification, structural characterisation and functional analysis of complexes important in RNA processing and translation in mammalian cells, are included in Chapter 5. Biophysical and genetic approaches including X-ray crystallography, NMR and three-hybrid assays are, however, beyond the scope of this book.

It is our intention that the book will provide scientists that are new in the RNA field with a comprehensive collection of thoroughly tested manuals, and those that are more experienced with a supplement to existing laboratory protocols.

Jørgen Kjems
Jan Egebjerg
Jan Christiansen

Contents

Chapter 3.   *Preparation of protein*   57

CHAPTER 1

Introduction to RNA-protein interactions

1.1. Introduction

The wide range of biological activities achieved by RNA molecules is reflected in a corresponding diversity of RNA structures. The three-dimensional structure of folded RNA is stabilized by a variety of interactions; the most prevalent are stacking and hydrogen bonding between bases. Stacking energy also accounts for different sugar and phosphate group interactions with bases.

Normally, RNA structures are described at three levels. The primary structure is the sequence including modifications. The secondary structure refers to interactions that can be depicted in two dimensions, whereas, tertiary interactions impose intersections of the RNA backbone, when depicted in two dimensions.

1.2. RNA secondary structures

1.2.1. Helices

The only differences between the chemical structures of RNA and DNA are a 2'-hydroxyl group in the former and a methyl group at the 5-position of uracil in the latter. The 2'-hydroxyl group restricts the structural flexibility of RNA, particularly in helices. DNA helices exhibit conformational flexibility, with the B-form as the prevalent form in solution, whereas, RNA helices are limited to A-forms. An important distinction between the A- and B-forms is the conformation (puckering) of the sugar moiety. In the A-form the sugar is in a 3'-*endo* puckering conformation where the 3'-carbon is placed on the same side as the 5'-carbon of the plane defined by the other four ring atoms. In the B-form, the sugar conformation is a 2'-*endo* puckering. The difference in puckering results in a significant change in the base pair tilt and a reduction in the rotation per nucleotide in the A-form. Therefore, the B-form helix is stabilized mainly by intrastrand base stacking while the A-form is stabilized by both inter- and intrastrand stacking interactions. The major macroscopic differences between the A- and B-form helices are the deep and narrow major groove and the shallow and wide minor groove of the A-form. This is a result of the displacement of the base pairs from the helical axis toward the minor groove and from the reduced phosphate-phosphate distance in the A-form.

The specificity of many DNA-protein interactions depends on hydrogen bonds between the protein (often an α-helix) and the functional groups of the bases in the major groove. The hydrogen-donating and hydrogen-accepting groups in the major groove can generate a unique pattern for recognition, while much less specificity is obtained for interactions in the minor groove. However, the narrow and deep major groove of a regular A-form helix cannot accommodate an α-helix. Protein interaction in the major groove can occur at the termini of the RNA helix where the large tilt of the bases makes the 3'-strand of the helix accessible. Hence, a regular A-form helix is, for sterical reasons, unlikely to be involved in specific protein interaction (Fig. 1.1).

1.2.2. Bulges and internal loops

Internal loops occur where two helices are separated by non-Watson-Crick base pairs. Bulges refer to cases where all bases on one strand can base pair while one or more bases on the opposite strand cannot (Fig. 1.1). Both bulges and internal loops are potential, and often used, targets for site-specific recognition of the RNA.

Internal loops are optimal structures for specific interactions. The

(a)

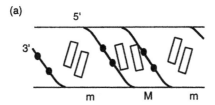

(b)

(c)

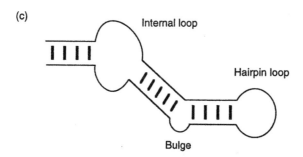

Fig. 1.1. Examples of two-dimensional representations of RNA structure elements. (a) View of an A-form helix from one side. The thick lines is drawn through the phosphate groups in the backbone. Groups which can be recognized from one site are indicated, i.e. groups which face the same side in the 3D structure: dots indicate phosphates, open bars indicate bases in minor groove (m) or major groove (M). (b) Conventional 2D presentation of a helix where the groups facing the same side in 3D are indicated as in (a). Notice the uneven stagger; 2–3 base pairs over minor groove while 6–7 over major groove. (c) Examples of unpaired segments. (d) A common representation of a pseudoknot. Notice that the orientations of helix A and B are different. (e, f) Same pseudoknot as (d) but both helices are drawn in the same orientation with either B stacked on A (e) or A stacked on B (f). Notice that a proper 2D depiction of a pseudoknot (e, f) results in crossing of the phosphate backbone, i.e. pseudoknots are tertiary structure elements.

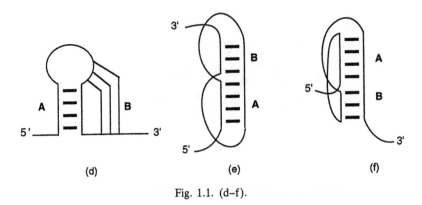

Fig. 1.1. (d–f).

bases in the loop will generally stack on each other and the flanking helices stabilize otherwise unstable non-Watson-Crick base pairs in the loop. The structural benefit is twofold: first, non-Watson-Crick base pairs create unique hydrogen bonding motifs, and second, the opening of the major groove increases the accessibility of the bases.

The bulged loops can either provide unique protein recognition sites by a direct interaction with the unpaired bases or by exposing the major groove at the termini of the adjacent helices. Moreover, a bulge may generate a bend in the RNA helix which can add to the specificity of a protein interaction.

1.2.3. *Hairpin loops*

A hairpin consists of a helix bridged by a loop of unpaired nucleotides (Fig. 1.1). The hairpin loop is a frequent target for protein interactions and also functions as nucleation site for RNA folding. In particular the 'tetraloop' hairpins, a group of loop sequences, UNCG, GNRA and CUUG, exhibit unusual thermodynamic stability.

1.2.4. Single-stranded regions

Single-stranded regions consist of unpaired nucleotides between helices in the secondary structure and at the 5' and 3' ends. The single-stranded regions may be ordered by base stacking in a helical geometry unless they are involved in tertiary structures.

1.3. RNA tertiary structures

The folding of RNA into a compact 3D structure is stabilized by tertiary interactions, which depend both on the nucleotide sequence and on the secondary structure. Although these interactions are structurally very diverse, some common features have been defined.

1.3.1. Triple-base pairs (single-strand/helix interactions)

Triple-base pairs occur when single-stranded nucleotides make hydrogen bonds with a Watson-Crick base pair. The third base may interact either within the major or the minor groove of the Watson-Crick base pair. For example, the folding of tRNA is stabilized by triple-base pairs between the junction loop and bases in the major groove of stem D. Also, a triple-base pair (UAU), between a U in a bulge and the U-A in the adjacent helix, is essential for the conformation of the binding site for human HIV protein Tat on TAR.

1.3.2. Pseudoknots (single-strand/single-strand interaction)

Originally, the terms 'knot' or 'pseudoknot' were used for any tertiary interaction between unpaired nucleotides in the secondary structure. However, recently the term 'pseudoknot' is used more specifically to describe Watson-Crick base pairing between bases in single-stranded loops and bases outside the loop (Fig. 1.1). The two helices stack coaxially and appear from one side as a continuous helix but from the other side two loops cross the major and the minor grooves. The number of nucleotides required to span the groove depend on the length of the helix. The close proximity of the negatively charged phosphates in the pseudoknot requires the presence of Mg^{2+} to shield the electrostatic repulsion.

1.3.3. Hairpin/internal loop interactions

More recent structural studies have revealed interaction between the minor groove of internal loops and hairpin loops. The GAAA subtype of the GNRA tetraloop can interact in the minor groove of a sequence motif [CCUAAG...UAUCC], called the tetraloop receptor. The interaction is very specific because each of the adenosines in the tetraloop makes specific interaction with the tetraloop receptor. A ribose zipper refers to an interaction between two backbones where the 2'-hydroxyl and the purine N3 (or pyrimidine O2) of a base in one strand interact with the 2'-hydroxyl of another strand. This motif might stabilize packing of the backbone in regions of tight helical packing.

1.4. Protein motifs involved in RNA binding

RNA binding proteins are structurally very diverse. Most of the ribosomal proteins are devoid of recognizable RNA binding motifs, whereas, the evolution of proteins involved with RNA processing, such as hnRNA, mRNA and snRNA binding proteins, has resulted in several fairly conserved RNA binding motifs (Table 1.1).

1.4.1. Ribonucleoprotein (RNP) motif

The RNP motif is also referred to as RNP consensus sequence (RNP-CS), RNA binding domain (RBD), or RNA recognition

TABLE 1.1
Examples of high resolution structures of RNA in complex with a ligand

RNP[1]
  U1A-U1 hairpin[2]
  U1A-U1A 3'UTR[3]
ARM
  BIV tat ARM-TAR[4]
  HIV tat ARM-TAR[5]
  Rev-RRE[6]
KH[7]
dsRBD[8], [9]
Others
  tRNA-synthetases[10]
  EF-Tu-tRNA-GTP[11]
  MS2 coat protein -RNA hairpin[12]
  ATP-RNA aptamer[13]
  Theophyllin-RNA aptamer[14]
  Citrullin/arginine-RNA aptamer[15]
  Antibiotic-RNA[16]

motif (RRM) and is the most common motif in RNA binding proteins. The RNP motif consists of 90–100 amino acids that are only moderately conserved with two recognizable regions; an octamer and a hexamer sequence called RNP1 and RNP2, respectively. RNP1 and RNP2 are separated by about 30 amino acids and characterized by a conserved distribution of phenylalanines and/or tyrosines. X-ray and NMR studies show that the RNP motif folds into a β_1-α_1-β_2-β_3-α_2-β_4 structure, where the β-stands form an antiparallel sheet, β_4-β_1-β_3-β_2, with the α-helices located on one side perpendicular to the β-sheet. The RNP1 and RNP2 are juxtapositioned in the three-dimensional structure on the β_3 and the β_1 strand, respectively.

The best characterized RNP motif in complex with RNA, is the U1A protein bound to its cognate small nuclear U1 RNA hairpin. U1A binds a 10-nucleotide hairpin structure. Different regions of the protein serve specialized functions in the interaction. The RNA hairpin is positioned on U1A by electrostatic interactions between the phosphate backbone of the RNA helix, and arginines and lysines in the loop between β_1 and α_1. The peptide loop between β_2 and β_3 prevents base pairing in the RNA loop by protruding through the RNA loop. Consequently, most of the bases in the loop are splayed out from the centre where the first seven bases at the 5′ side of the loop fit into the groove between the β_2-β_3 loop and the C terminus of the domain. The specific interaction is achieved in the single-stranded region where bases are stacked on aromatic side chains in addition to many direct and water-mediated hydrogen bonds.

1.4.2. The K homology domain (KH)

The KH domain was originally identified in hnRNP K (hnRNP-K-Homologous). It is approximately 50 amino acids containing an octapeptide IGXXGXXI (X can be any amino acid) where the GXXG part is the most conserved. The KH domain is present in single or multiple copies in various RNA-binding proteins. Recent studies have revealed the 3D structure of a KH domain in FMR1. The domain folds as a β_1-α_1-(F)-β_2-β_3-α_2 structure where the 15 amino acids region (F) between α_1 and β_2, containing the octapeptide, is very flexible in the free protein. The β-strands fold in a sheet topology β_1-β_3-β_2 with an exposed surface of hydrophilic residues on one side and a hydrophobic surface supported by the amphipatic α_2 helix on the other side. The KH domain binds homopolymers, particularly polypyrimidines, but no *in vivo* targets have yet been identified for direct interaction with the KH domain. The RNA binding region has not been identified but mutagenesis data suggest that the flexible region might be involved.

1.4.3. The arginine-rich motif (ARM)

In the analysis of nucleic acid protein interactions, it is generally advantageous to distinguish between interactions which define affinity and specificity. Affinity and specificity are sometimes erroneously equated but the affinity can be increased by fairly unspecific electrostatic interactions between positively charged amino acids and the phosphate backbone. The increase in the number of such interactions will result in increased affinity but, most likely, reduced specificity. The distinction between interactions which contribute to affinity or specificity is relevant for the arginine-rich motif (ARM) which is composed of a 8–20 amino acids motif particularly rich in arginine residues.

The interaction between the HIV protein Tat and its cognate *trans*-activation response element TAR involves an arginine-rich motif. Substituting this RNA-binding ARM region (RKKRRQRRR) in tat with homolysine reduces both affinity and specificity for TAR. However, replacing one arginine into the homolysine stretch at a position with three or four lysines on each side recovers the specific binding to TAR. The guanidinium moiety of a single arginine imposes the specificity by binding and stabilizing a cavity formed in the TAR by a specific triple-base (see above). The interaction is referred to as an arginine fork. Arginine alone, although in the millimolar range, can bind TAR specifically. The 10^6-fold higher affinity of the ARM sequence is contributed by electrostatic interactions between the flanking basic residues and the phosphate backbone.

Apart from arginine, there is little similarity between different ARMs. The two other known structures of ARM peptides in complex with their respective RNAs fold differently. The rev ARM interacts as an α-helix in the internal loop of RRE structures, while the ARM of the bovine tat binds TAR in a β-turn like structure.

1.4.4. The RGG box

The RGG domain was initially identified in hnRNP U and is often found in several copies in RNA binding proteins. It consists of a 20–25 amino acids domain containing several RGG tripeptides often separated by aromatic residues. Little is known about the structure of the RGG domains and RNA specificity. In some cases, as for hnRNP U, both the specificity and affinity reside in the RGG domain while for nucleolin the RGG domains are important for high RNA affinity, and the determinant for specific interaction resides outside the RGG domain. The hydrogen bonding properties of the guanidinium moiety of arginine may contribute to the RNA binding. Interestingly, many of the arginines in RGG domains are modified to N^G, N^G-dimethylarginine, which may sterically prevent RNA binding and thereby serve as a potential modulatory mechanism.

1.4.5. Zinc fingers

The zinc-finger motif was originally identified in *Xenopus* transcription factor IIIA (TFIIIA). TFIIIA has the unique property of binding specifically to both 5S rRNA and the 5S rRNA gene. In the TFIIIA zinc-fingers, Zn^{2+} is coordinated by a pair of cysteines and a pair of histidines (called the C_2H_2 type) and the intervening 12–14 amino acids fold in an antiparallel β-sheet and an α-helical segment. Numerous zinc finger proteins of the C_2H_2 type have been identified but only a few of these have been shown to bind RNA. The generality of zinc finger-RNA binding remains to be determined.

1.4.6. Double-strand RNA binding domains (dsRBD)

The double-strand RNA binding domain is a 70 amino acids globular domain that folds in α_1-β_1-β_2-β_3-α_2 structure. The two α-helices lie on one side of a three-stranded antiparallel β-sheet. dsRBD binds exclusively double-stranded RNA in the minor groove by recognition of the 2'-hydroxyl groups. Isolated dsRBD binds approximately 16 base pair helical RNA nonspecifically and more specific binding is presumably established by the concurrent binding of two or several dsRBD present in most dsRBD containing proteins. Model building suggests that amino acids in α_1 and in the linkers between β_1 and β_2 and between β_3 and α_2 interact with the 2'-hydroxyl groups.

References

1. Nagai, K., Oubridge, G., Jessen, T.H., Li, J. and Evans, P.R. (1990). Crystal structure of the RNA-binding domain of the U1 small nuclear ribonucleoprotein A. Nature *348*, 515–520.
2. Oubridge, C., Ito, N., Evans, P.R., Teo, C.H. and Nagai, K. (1994). Crystal structure at 1.92 Å resolution of the RNA-binding domain of the U1A spliceosomal protein complexed with an RNA hairpin. Nature *372*, 432–438.
3. Allain, F.H.-T., Gubser, C.C., Howe, P.W.A., Nagai, K., Neuhaus, D., and Varani, G. (1996). Specificity of ribonucleoprotein interaction determined by RNA folding during complex formation. Nature *380*, 646–650.
4. Puglisi, J.D., Chen, L., Blanchard, S. and Frankel, A.D. (1995). Solution structure of a bovine immunodeficiency virus Tat-TAR peptide-RNA complex. Science *270*, 1200–1203.
5. Puglisi, J.D., Tan, R., Calnan, B.J., Frankel, A.D. and Williamson, J.R. (1992). Conformation of the TAR RNA-arginine complex by NMR spectroscopy. Science *257*, 76–80.
6. Battiste, J.L., Mao, H., Rao, S., Tan, R., Muhandiram, D.R., Kay, L.E., Frankel, A.D., and Williamson, J.R. (1996). α helix-RNA major groove recognition in an HIV-1 Rev peptide-RRE RNA complex. Science *273*, 1547–1551.
7. Musco, G., Kharrat, A., Stier, G., Fraternali, F., Gibson, T.J., Nilges, M., and Pasore, A. (1997). The solution structure of the first KH domain of FMR1, the protein responsible for the fragile X syndrome. Nat. Struct. Biol. *4*, 712–716.
8. Kharrat, A., Macias, M.J., Gibson, T.J., Nilges, M. and Pastore, A. (1995). Structure of the dsRNA binding domain of E. coli RNase III. EMBO J. *14*, 3572–3584.
9. Bycroft, M., Grünert, S., Murzin, A.G., Proctor, M. and Johnston, D.S. (1995). NMR solution structure of a dsRNA binding domain from Drosophilia staufen protein reveals homology to the N-terminal domain of ribosomal protein S5. EMBO J. *14*, 3563–3571.
10. Cavarelli, J. and Moras, D. (1993). Recognition of tRNAs by aminoacyl-tRNA synthetases. FASEB J. *7*, 79–86.
11. Nissen, P., Kjeldgaard, M., Thirup, S., Polekhina, G., Reshetnikova, L., Clark, B.F.C., and Nyborg, J. (1995). Crystal structure of the ternary complex of Phe-tRNAphe, EF-Tu and a GTP analog. Science *270*, 1464–1472.
12. Valegard, K., Murray, J.B., Stonehouse, N.J., van den Worm, S.,

Stockley, P.G., and Liljas, L. (1997). The three-dimensional structures of two complexes between recombinant MS2 capsids and RNA operator fragments reveal sequence-specific protein- RNA interactions. J. Mol. Biol. *270*, 724–738.

13. Dieckmann, T., Suzuki, E., Nakamura, G.K. and Feigon, J. (1996). Solution structure of an ATP-binding RNA aptamer reveals a novel fold. RNA *2*, 628–640.

14. Zimmermann, G.R., Jenison, R.D., Wick, C.L., Simorre, J.P. and Pardi, A. (1997). Interlocking structural motifs mediate molecular discrimination by a theophylline-binding RNA. Nat. Struct. Biol. *4*, 644–649.

15. Yang, Y., Kochoyan, M., Burgstaller, P., Westhof, E. and Famulok, M. (1996). Structural basis of ligand discrimination by two related RNA aptamers resolved by NMR spectroscopy. Science *272*, 1343–1347.

16. Fourmy, D., Recht, M.I., Blanchard, S.C. and Puglisi, J.D. (1996). Structure of the A site of Escherichia coli 16S ribosomal RNA complexed with an aminoglycoside antibiotic. Science *274*, 1367–1371.

CHAPTER 2

Preparation of RNA

2.1. Working with RNA

2.1.1. Special precautions when working with RNA

There are four parameters which must be minimised when working with RNA: Basic pH, high temperature, heavy metal ions and ribonucleases. The first three parameters are kept under control by working at neutral or slightly acidic pH at 0°C in the presence of 0.1 mM EDTA. Obviously these conditions cannot always be met due to other factors of importance in RNA-protein studies such as pH and temperature optima of enzymes (e.g. AMV reverse transcriptase and alkaline phosphatease), keeping acidic proteases at bay during cell lysis, or the need to preserve a native RNA conformation during complex formation and isolation of endogenous ribonucleoprotein particles.

The unwanted presence of ribonucleases is usually regarded as the most critical parameter when working with RNA, and there is a considerable amount of lablore in connection with this issue. However, in the majority of experiments the source of ribonucleases should only be the cells from which RNA or ribonucleoprotein particles are extracted. Therefore, disposable plasticware should be used instead of glassware from the departmental washing-up facility. However, glassware can be treated with a mixture of chromic and sulfuric acids followed by a rinse with EDTA-containing H_2O, and nondisposable plasticware can be treated with 0.1 M NaOH, 1 mM EDTA and rinsed with H_2O. All solutions should either be autoclaved or steril-filtered to prevent microbial growth. However,

13

we find the prevalent use of diethylpyrocarbonate-treated H_2O superfluous, and potentially problematic, when it is recalled that this reagent modifies adenosines. Current protocols for plasmid minipreps, UV-crosslinking experiments, RNase protection experiments, and *in vitro* translation studies, often include a step where RNase A is used to degrade total RNA. Generally, we avoid these protocols and use the less aggressive RNase T_1, sometimes supplemented with the double-strand specific RNase V_1.

Release of endogenous ribonucleases during cell rupture is seldom a problem when lysis is carried out in the presence of a denaturing reagent, such as guanidinium thiocyanate, but certain tissues, such as placenta and pancreas, are notorious for giving rise to degraded RNA if the denaturation is not fast enough. The most difficult type of experiment in terms of preserving RNA integrity is that where cell lysis has to be carried out without jeopardising endogenous RNA-protein complexes. In this case, the use of a ribonuclease inhibitor, such as recombinant RNasin or heparin is called for, whereas, the use of these inhibitors in other circumstances ought to be unnecessary. The recombinant form of RNasin rather than the placental extract is recommended for use, since the latter contains large amounts of the ribonuclease angiogenin that is released from RNasin by heat.

2.1.2. Basic protocols for RNA work

This section contains descriptions of basic procedures that occur throughout the book. In this way, we hope to eliminate the repetition of basics such as RNA quantification, gel extraction, extraction with organic solvents, RNA precipitation, long-term storage of RNA, renaturation, the use of spin-columns and preparation of standard reagents.

2.1.2.1. RNA quantification
Absorption of ultraviolet light at 260 nm is used to assess the amount of RNA in a sample, since 1 A_{260} unit equals an RNA

concentration of 40 µg/ml. Obviously, the sample must not contain interfering aromatics such as phenol, and the presence of millimolar amounts of compounds such as 2-mercaptoethanol and dithiothreitol will also lead to erroneous results. It is customary also to carry out absorbance readings at 230 and 280 nm to estimate the interference by polysaccharides and proteins, respectively. The optimal ratio of optical density for an RNA with average composition at 230, 260 and 280 nm is 1 : 2 : 1 provided pH is slightly basic (8.0–8.5). The major drawbacks of the spectrophotometric assessment of RNA concentration in a standard spectrophotometer is the need to use about 4 µg of precious RNA to get a reliable reading from a 1-ml cuvette and the often cumbersome adjustment of wavelengths. Therefore, spectrophotometric equipment designed especially for nucleic acid quantification has been produced, and the GeneQuant apparatus from Pharmacia Biotech can use a 7-µl cuvette and measure absorbance at the three common wavelenghts in a straightforward manner.

An alternative to spectrophotometric quantification is gel electrophretic analysis of the RNA sample in parallel with dilutions of a known RNA standard. Following staining with 0.1% toluidine blue in 7.5% acetic acid for 15 min and destaining with 2.5% acetic acid, the amount in the sample can be estimated. Although the gel electrophoretic procedure is more tedious than the spectrophotometric approach, it has the inherent advantage of establishing the integrity of the isolated material.

2.1.2.2. Gel purification
Full-length RNA molecules up to about 500 nucleotides can be purified by denaturing polyacrylamide gel electrophoresis. Following electrophoresis, RNA is localised in the gel by UV_{254}-shadowing over Xerox paper, and a gel slice containing the full-length molecule is excised. The gel slice is crushed and continuously shaken at room temperature with equal volumes of 0.25 M sodium acetate (pH 6.0), 1 mM EDTA and phenol, followed by chloroform extraction and precipitation with ethanol (see below for further details).

The necessary time of shaking will depend on the size of RNA, but overnight elution of reasonable amounts of large RNAs is feasible.

Alternatively, large RNA molecules can be electroeluted from the gel slice into a dialysis bag containing TBE buffer by placing the dialysis bag perpendicularly to the voltage gradient in an electrophoresis chamber with TBE and electrophoresing at 10 V/cm for 1 h.

2.1.2.3. Extraction with organic solvents

A standard deproteinisation includes one extraction with an equal volume of phenol saturated with TE, one extraction with a 1:1 mixture of TE-saturated phenol and chloroform, and finally one extraction with chloroform. Samples are centrifuged at 5000 g for 5 min, and the upper phase is transferred to a new tube between each extraction.

2.1.2.4. Precipitation

RNA is precipitated from solution by adding one-tenth the volume of 2.5 M sodium acetate (pH 6.0) and 2.5 times the aqueous volume of ethanol. Mix well and leave on dry ice for 5 min, followed by centrifugation at 10,000 g for 15 min at 4°C. If a small amount of RNA (less than 0.1 µg/ml) is going to be recovered, inclusion of either 20 µg/ml *E. coli* tRNA or glycogen as a carrier is recommended. Wash the precipitate with 70% ethanol and centrifuge at 10,000 g for 5 min. Let remaining traces of ethanol evaporate (or use a speed-vac briefly), before the precipitate is dissolved in double-distilled H_2O or TE buffer.

High-molecular-weight RNAs can be precipitated without the concomitant precipitation of DNA by the addition of one-tenth volume of 8 M LiCl and leaving the sample at 0°C for at least 2 h, before centrifugation at 15,000 g for 20 min at 4°C.

2.1.2.5. Storage of RNA

It is convenient to store RNA at a concentration of about 5 μg/μl in slightly acidic double-distilled H_2O at –80°C, since this provides flexibility in terms of possible options. For short-term storage, –20°C will suffice, and TE buffer is an alternative to double-distilled H_2O.

2.1.2.6. RNA renaturation

Since so much RNA biochemistry is carried out on RNA that has been purified in the presence of various denaturants such as guanidinium thiocyanate, urea, formamide, chelators, phenol and even distilled water, the issue of renaturation is crucial. One view is that the presence of these compounds should be avoided altogether, since these treatments denature full-length molecules which at later stages may be trapped in nonnative conformations. Although this view is accommodated with cotranscriptionally folded *in vitro* transcripts, it is difficult to isolate intact endogenous RNA from a cellular environment without the use of denaturants. Therefore, the conventional approach is to employ a wide range of denaturants during RNA preparation and then apply an RNA renaturation protocol, that ideally should result in the formation of a single native conformer. The importance of working with a conformationally homogeneous population should not be underestimated in structure-function studies, since probing data from a mixture of forms are uninformative, and nonnative forms may out-titrate *trans*-acting factors and, thus, invalidating functional assays.

Here we present an RNA renaturation protocol that is useful for both small and large RNA molecules. The important feature of the protocol is that secondary structure formation is favoured before tertiary structure formation by introducing Mg^{2+} at a late stage in the protocol:

1. Heat purified RNA in 20 mM Tris-HCl, pH 7.8, 140 mM KCl at 90°C for 1 min.
2. Transfer to 60°C and leave for 15 min.
3. Cool slowly to 30°C over a 15-min period.

4. Add $MgCl_2$ to a final concentration of 2.5 mM and leave at 30°C for 15 min.
5. Transfer to 0°C.

Renatured RNAs can be examined for gross conformational homogeneity by nondenaturing gel electrophoresis in either polyacrylamide gels or Nusieve agarose gels. Electrophoresis is carried out in TBE supplemented with 5 mM $MgCl_2$ and 50 mM KCl at 4 V/cm at room temperature.

2.1.2.7. Desalting and removal of nucleotides

Change of buffer and removal of nucleotides are conveniently carried out by spin-column chromatography. One can either use homemade gel filtration matrix-based columns in 1 ml syringes, or purchase ready-to-use cartridges (e.g. MicroSpin™ Columns from Pharmacia Biotech). In both cases, a sample volume of 100 µl can be handled after the matrix has been drained by a prespin.

2.1.2.8. Diethylpyrocarbonate

As mentioned in a previous section, we generally find the use of diethylpyrocarbonate as a ribonuclease inhibitor superfluous. However, if glassware is shared with labs that use RNase A in their solutions, problems may arise. Therefore, beakers, tubes, etc. that are to be used for RNA work can be filled with a 0.1% (v/v) of diethylpyrocarbonate in water and left for a couple of hours at 37°C in a fume-hood. Rinse thoroughly with distilled water, and remove the final traces of diethylpyrocarbonate by autoclaving.

2.1.2.9. Standard reagents

More specialised reagents are described in each section.

Buffers: Common buffers are Tris (pK_a 8.1) and HEPES (pK_a 7.5) that in many cases are interchangeable. Both buffers can be made as 1 M stocks solutions adjusted with HCl and KOH, respectively. However, Tris buffers exhibit a high degree of temperature sensitivity (−0.03 pH/°C), and due to the nucleophilicity of the base,

Tris buffers are incompatible with the presence of several probing reagents.

rNTPs: Dissolve each ribonucleoside triphosphate in H_2O at a concentration of 50–100 mM. Titrate each solution to pH 7.5 with 1 M Tris base using indicator strips. The exact concentration of each neutralised ribonucleoside triphosphate solution can now be estimated by an absorbance measurement (ϵ_{259} (A) = 1.54×10^4 M^{-1} cm^{-1}; ϵ_{253} (G) = 1.37×10^4 M^{-1} cm^{-1}; ϵ_{271} (C) = 9.1×10^3 M^{-1} cm^{-1}; ϵ_{262} (U) = 1.0×10^4 M^{-1} cm^{-1}). Prepare an rNTP stock solution containing 10 mM of each ribonucleoside triphosphate (neutralised solutions of each ribonucleoside triphosphate are also commercially available).

2.5 M sodium acetate (pH 6.0): Dissolve 20.5 g anhydrous sodium acetate in about 90 ml H_2O and adjust pH to 6.0 with acetic acid before the final volume is made up to 100 ml (strictly speaking this is an incorrect way of generating a 2.5 M buffer but it ensures that the Na^+ concentration is 2.5 M; the correct way would be to titrate 2.5 M sodium acetate with 2.5 M acetic acid to obtain the desired pH).

TE: 10 mM Tris-HCl, pH 7.5, 0.1 mM EDTA

TBE: 89 mM Tris-borate, pH 8.3, 2 mM EDTA

tRNA carrier: RNase free (Boehringer), 5 mg/ml stock solution.

Phenol: Purest grade. Store at 4°C

Phenol/Chloroform: (1:1)

2.2. Isolation of RNA from cells

Isolation of RNA from cells or tissues is usually carried out in the presence of the chaotropic agent guanidinium thiocyanate with the purpose of disrupting the cellular structures with concomitant inhibition of ribonucleases. In this section, two basic procedures that use guanidinium thiocyanate are presented. The first is a modified version of the original procedure of Chirgwin et al.[1] that is appropriate for handling a small number of samples on a reasonably large scale (grams of tissues or about 10^8 cells). The second procedure, that in various guises forms the basis of RNA isolation kits from commercial sources, is the single-step guanidinium thiocyanate acid-phenol method originally developed by Chomczynski and Sacchi,[2] which is suitable for handling a large number of samples on a small scale. Moreover, a hot phenol—SDS method[3] useful for isolating RNA from sources, such as yeasts, plants and bacteria, is also presented, as is a vanadyl-ribonucleoside complex method[4] for preparation of nuclear and cytoplasmic RNA. The section ends with a slightly modified version of the standard procedure of obtaining polyadenylated RNA.[5]

2.2.1. Guanidinium thiocyanate–CsCl method

Materials
CsCl cushion
    5.7 M CsCl (Merck)
    100 mM EDTA, pH 7.4
    filtered and autoclaved
Guanidinium thiocyanate lysis solution
    Dissolve 250 g guanidinium thiocyanate (Fluka) in 21 ml 1 M Tris-HCl, pH 7.6, 8.5 ml 10 mM EDTA and 150 ml H_2O by stirring for 3–4 h. The solution is then filtered and 21 ml 2-mercaptoethanol (Merck) is added. Finally, pH is adjusted to 7.0 with NaOH and the volume is made up to 420 ml.
Proteinase K solution
    200 µg/ml Proteinase K (Pharmacia, Promega)
    20 mM Tris-HCl, pH 7.5
    2 mM EDTA
    0.2% SDS
    Autodigest for 1 h at 37°C before use
Sodium N-lauroylsarcosinate (Fluka)
CsCl 5.7 M

Equipment
    50 ml polypropylene tubes (Falcon, Nunc)
    Polytron
    Ultra-Clear^{TM} tubes (14 × 89 mm, Beckman)
    Ultracentrifuge and SW41 rotor

Procedure
1. Adherent cells (about 10^8) can be lysed directly in the culture flasks whereas cells in suspension culture must be pelletted. Fresh tissue (about 1 g) should be cut into small pieces, and frozen tissue and material of plant origin ought to be crushed under liquid nitrogen.[a]
2. Transfer the sample to a 50 ml polypropylene tube and add 7.5 ml *guanidinium thiocyanate lysis solution*.
3. Homogenise at full-speed for 1 min with a Polytron.
4. Add 0.35 g *sodium N-lauroylsarcosinate*. Cap and invert the tube a few times—leave for 5–10 min.
5. Add 3.7 ml *5.7 M CsCl* and mix.[b]
6. Place 2 ml *CsCl cushion* at the bottom of a polyallomer ultracentrifuge tube (14 × 89 mm).
7. Layer the homogenate carefully on top of the cushion.
8. Centrifuge at 100,000 g (r_{max}) for 20 h at 20°C in a SW41 rotor (24,000 rpm).[c]
9. Remove the majority of the supernatant with a 10 ml disposable pipette taking care not to touch the bottom of the tube. Pour off the remainder of the supernatant and leave the tube in an inverted position.
10. Cut off the bottom of the tube with a heated scalpel.
11. Resuspend the clear RNA pellet in 200 µl autodigested[d] *proteinase K solution* and transfer to a microfuge tube. Wash the polyallomer tube bottom with an additional 200 µl of *proteinase K solution* and combine in the microfuge tube.
12. Leave for 1 h at 37°C.
13. Extract with 2 × 300 µl phenol/chloroform and 300 µl chloro-

form transferring the aqueous phase to a new microfuge tube after each extraction.

14. Add 1/10 volume 2.5 M sodium acetate (pH 6) and 2.5 volumes ethanol, mix and leave for 20 min on dry ice.
15. Centrifuge at 10,000 g for 15 min at 4°C.
16. Remove supernatant and wash the RNA pellet with 70% ethanol.
17. Centrifuge at 10,000 g for 5 min at 4°C.
18. Remove supernatant and dry the pellet briefly in a vacuum centrifuge.
19. Redissolve the dried pellet in 100 µl double-distilled and autoclaved H_2O.
20. Estimate the amount and the purity by measuring both the A_{260} and the A_{280}.

Notes

a. Use eye-protection during the crushing and do not allow the pieces to thaw. Transfer the tissue pieces in liquid nitrogen into the polypropylene tube, and allow the liquid nitrogen to evaporate. Then add the lysis solution immediately.
b. If insoluble material from tissue samples is present at this stage it should be removed by filtering through a Whatman 54 filter.
c. The temperature should not be lower, since this may lead to precipitation of CsCl in the cushion.
d. It is crucial that the proteinase K solution is autodigested by incubation at 37°C after it has been prepared, so traces of ribonucleases are removed.

Comments

The guanidinium thiocyanate—CsCl procedure is a lengthy process but it generates high-quality RNA with a A_{260}/A_{280} ratio between 1.8 and 2.0. The integrity of the isolated RNA can be judged from denaturing agarose gel electrophoresis followed by ethidium bromide staining, since the 28 S ribosomal RNA band ought to exhibit a considerably greater fluoresecens than that obtained from the 18 S ribosomal RNA band. Nevertheless, placenta and pancreas with their high ribonuclease contents are difficult to work with, and liver contains large amounts of glycogen that may lead to a slight contamination of the RNA preparation. A critical step in the procedure is the redissolution of the RNA pellet after ultracentrifugation, since the denaturants are absent at this stage. Therefore, the inclusion of an autodigested proteinase K solution provides a ribonuclease-free environment during this step and has the added advantage of degrading any proteinaceous material in the pellet. The glass-clear RNA pellet at the bottom of the ultracentrifugation tube is sometimes difficult to see, but after the proteinase K solution has been added the release of the pellet can be observed. Although cells are easily disrupted in the guanidinium thiocyanate lysis solution the mechanical homogenisation is necessary to shear the DNA, so the RNA is not trapped in high-molecular weight DNA during the ultracentrifugation step. The guanidinium thiocyanate–CsCl procedure is generally considered as a method for obtaining total RNA, but small RNAs such as tRNAs, 5S RNA, snRNAs and snoRNAs are not recovered quantitatively.

2.2.2. Single-step guanidinium thiocyanate acid-phenol method

Materials
Denaturing solution
   4 M guanidinium thiocyanate (Fluka)
   25 mM sodium citrate, pH 7.0
   0.5% N-lauroyl-sarcosinate (Fluka)
   100 mM 2-mercaptoethanol
Sodium acetate (2 M), pH 4.0
Phenol (H_2O-saturated)
Chloroform-isoamyl alcohol (49:1)
Isopropanol

Equipment
Hand-held homogeniser
5 or 10 ml polypropylene tubes

Procedure

1. Homogenise 100 mg tissue or 10^7 cells[a] in 1 ml *denaturing solution* and transfer to a 5-ml polypropylene tube.
2. Add 100 μl 2 M *sodium acetate, pH 4.0* and mix by inversion.
3. Add 1 ml *phenol (H_2O-saturated)*[b] and mix by inversion.
4. Add 200 μl *chloroform-isoamyl alcohol (49:1)* and vortex for 10 s.
5. Leave on ice for 15 min and centrifuge at 10,000 *g* for 10 min at 4°C.
6. Avoid the interphase and transfer the aqueous phase[c] to a 2 ml microfuge tube.
7. Add an equal volume of *isopropanol*, mix and place at –20°C for 1 h.
8. Centrifuge at 10,000 *g* for 20 min at 4°C and discard the supernatant.
9. Redissolve the RNA pellet in 300 μl *denaturing solution* and transfer to a new 1.5 ml microfuge tube.
10. Precipitate with 300 μl isopropanol at –20°C for 1 h.
11. Centrifuge at 10,000 *g* for 20 min at 4°C and discard the supernatant.
12. Redissolve the RNA pellet in 200 μl 0.25 M sodium acetate, pH 6.0 and ethanol-precipitate.[d]
13. Dissolve the dried pellet in 20 μl H_2O.

Notes

a. Use a hand-held homogeniser, or in the case of cells the lysis solution may be passed several times through a 23G-needle in the presence of 0.1% antifoam A.
b. Do not use neutralised phenol.
c. The volume is unlikely to exceed 1 ml.
d. The ethanol precipitation step is introduced to remove traces of guanidinium thiocyanate.

Comments

Since the guanidinium thiocyanate acid-phenol method can process a large number of samples on a small scale, several manufacturers base their total RNA isolation kits on this method. Therefore, efforts have been directed towards simplifying the original procedure. The major short-cut is to incorporate phenol in the lysis solution, but the kit protocols tend to stick with the original ratio of 1 ml lysis solution per 100 mg tissue or 10^7 cells. However, up to 10^8 cells from leucocytic cell lines have been homogenised in 600 µl lysis solution, but the resulting highly viscous lysate was then sheared by passing through a thin needle and the RNA precipitated by 4 M LiCl prior to the acid-phenol extraction.[6] Since all protocols include isopropanol precipitation steps, minute RNA samples can only be recovered by adding a carrier such as 20 µg/ml glycogen. However, recent RNA binding resin procedures (e.g. BIO 101) circumvent the precipitation steps, so small amounts of RNA can be recovered without the use of a carrier.

2.2.3. *The hot phenol–SDS method*

This is a popular method for preparation of RNA from bacterial, archaeal and small eukaryotes (e.g. yeast). The method exploits the property of DNA to go into the phenol phase together with the proteins at pH 5.0.

Materials
Acetate buffer
  50 mM sodium acetate, pH 5.0
  10 mM EDTA
Sodium acetate 3.0 M, pH 5.3
SDS 10%
Phenol (saturated with acetate buffer)

Equipment
50 ml polypropylene tubes
Waterbath at 65°C

Heating block at 65°C
Dry ice-ethanol bath

Procedure
1. Transfer 10 ml bacterial or yeast cultures to a 50 ml polypropylene tube and centrifuge at 3000 rpm for 5 min at 4°C.[a]
2. Discard the supernatant and wash the pellet in 1 ml *acetate buffer*. Transfer to a microfuge tube and centrifuge for 3 min at 4°C.
3. Discard the supernatant and resuspend the pellet in 500 μl ice-cold *acetate buffer*.
4. Add 50 μl *10% SDS* and mix briefly.
5. Add 550 μl *phenol* (prewarmed at 65°C), vortex for 1 min,[b] and place in a heating block for 5 min at 65°C (vortex briefly every minute).[c]
6. Chill quickly to room temperature in a water bath and centrifuge for 5 min at room temperature.
7. Discard the organic phase and leave the interphase and debris in the tube.
8. Repeat steps 5–6.
9. Transfer the upper aqueous phase to a new tube, taking care to avoid the interphase.
10. Add 400 μl chloroform and extract vigorously. Centrifuge for 5 min.
11. Transfer the aqueous layer to a new microfuge tube. Add 0.1 vol 3 M *sodium acetate* and precipitate with 2.5 vol ethanol.[d]

Notes
a. Tissue of plant origin will have to be crushed in a mortar under liquid nitrogen before it is transferred to the polypropylene tube (see *Note a* in Section 2.2.1 and *Note b* below).
b. The yield of extracted RNA can be improved by including 2 g acid-washed glass beads.
c. A combined heating block and shaker such as the Eppendorf 5436 thermomixer is useful.

d. The isolated RNA can be further purified by selective precipitation from 2 M LiCl (Section 2.1.2.4) followed by an additional ethanol precipitation. However, the LiCl precipitation will jeopardise the yield of RNAs <200 nucleotides.

2.2.4. Vanadyl-ribonucleoside complex method for preparation of nuclear and cytoplasmic RNA

Materials
PBS
  20 mM potassium phosphate, pH 7.4
  0.13 M NaCl
Lysis buffer
  20 mM HEPES-KOH, pH 7.2
  140 mM KCl
  1.5 mM $MgCl_2$
  0.5% Nonidet P-40
  10 mM vanadyl-ribonucleoside complex (Gibco-BRL)
Sucrose cushion
  20 mM HEPES-KOH, pH 7.2
  140 mM NaCl
  1.5 mM $MgCl_2$
  1% Nonidet P-40
  24% (w/v) sucrose
Proteinase K solution (Autodigest for 1 h at 37°C before use)
  0.3 M Tris-HCl, pH 7.5
  40 mM EDTA
  0.5 M NaCl
  3% w/v SDS
  0.5 mg/ml proteinase K (Pharmacia, Promega)
DNase I solution
  60 mM Tris-HCl, pH 7.5
  10 mM $MgCl_2$
  1 mM EDTA
  10 units DNase I (RNase-free from Pharmacia).

Extraction buffer
  0.3 M sodium acetate, pH 6.0
  15 mM EDTA
  0.3%   SDS.

Equipment

Swinging bucket microfuge

Procedure

1. Rinse the dish (100 mm) twice with 10 ml ice-cold *PBS*.
2. Add 1 ml of *PBS* containing 2 mM EDTA, wait 30 s, and then scrape off cells into an Eppendorf tube. Spin for 10 s at 12,000 rpm and wash pellet with 1 ml *PBS*.
3. Resuspend cells in 300 µl *lysis buffer*. Vortex gently for 10 s and leave on ice for 1 min.
4. Carefully underlay the lysate with 200 µl of *sucrose cushion*. Spin at 7500 *g* for 15 min in swinging buckets.
5. Collect 200–250 µl of the turbid upper layer (cytoplasmic fraction) and add 1/2 volume of *proteinase K solution*. Incubate at 37°C for 30 min, extract with phenol/chloroform twice, and precipitate with ethanol.[a]
6. Resuspend pelleted nuclei in 300 µl *lysis buffer* and add 1/2 volume of *proteinase K solution* **without** proteinase K. Sonicate to shear chromosomal DNA and add proteinase K to a final concentration of 0.2 mg/ml. Digest for 30 min at 37°C.
7. Add 50 µl of *DNase I solution* to the nuclear fraction and incubate at 37°C for 1 h.
8. Add 150 µl of *extraction buffer* and extract twice with phenol-chloroform, once with chloroform and precipitate with ethanol.

Notes

a. If working with transfected cells, omit the extractions with organic solvents and the ethanol precipitation and proceed to steps 7 and 8. Incubate for 30 min at 37°C.

2.2.5. Purification of poly-A$^+$ RNA

Although the function of the poly(A)-tail on mRNAs is still an unsettled issue, this post-transcriptional modification has provided a useful handle for purifying mRNAs. Virtually all procedures are based on the original study by Aviv and Leder[5] that used affinity chromatography with oligo(dT)-cellulose to isolate the less than 5% of total RNA that is mRNA. The general principle is that an RNA-DNA hybrid is formed between the poly(A)-tail and the oligo(dT)-cellulose at 500 mM salt, and that excess of rRNAs, tRNAs and other small RNAs can be washed away at this ionic strength. The captured mRNAs is then eluted by removing the salt, thus disrupting the base-pairing. The method is reliable and has stood the test of time. Although the oligo(dT)-cellulose is rather expensive, it has excellent capacity and can be reused after washing with 0.1 M NaOH. The major drawback of the method is the need to start with total RNA on a mg scale to avoid substantial losses during chromatography, extraction with organics and precipitations.

Due to these limitations several manufacturers are now offering kits whereby poly-A$^+$ RNA can be isolated directly from cell lysates on a small scale bypassing isolation of total RNA. Although the kits are based on the same general principle as outlined above, they can be divided into two types: (1) Small oligo-dT-containing latex or cellulose beads that are able to stay in a fine suspension during hybridisation and is pelleted by centrifugation; and (2) Paramagnetic particles that either contain covalently attached oligo(dT) or streptavidin. In the latter case, the hybridisation step is carried out in solution between poly-A$^+$ RNA and biotin-tagged oligo(dT), and the hybrids are collected on the streptavidin-coated paramagnetic beads. Note that the matrix in many of the kits cannot be reused.

In the following a slightly modified version of the conventional oligo(dT)-cellulose chromatography procedure suitable for obtaining poly-A$^+$ RNA from about 1–10 mg total RNA is presented.

Materials
Oligo(dT)-cellulose (Gibco-BRL, Pharmacia)
2 × load buffer
  20 mM Tris-HCl, pH 7.5
  800 mM NaCl
  2 mM EDTA
  0.2% SDS
Medium-salt buffer
  10 mM Tris-HCl, pH 7.5
  100 mM NaCl
  1 mM EDTA
  0.1% SDS
Elution buffer
  10 mM Tris-HCl, pH 7.5
  1 mM EDTA
  0.1% SDS
Wash buffer
  0.1 M NaOH
  5 mM EDTA

Equipment
Small plastic chromatography columns (Bio-Rad)

Procedure
1. Suspend *0.2 g oligo(dT)-cellulose powder* in 2 ml *wash buffer* and pour into the column.[a]
2. Wash with 15 ml *wash buffer*.
3. Rinse the packed column with 15 ml H_2O or until pH of the effluent is 8.5.
4. Equilibrate the column with 10 ml *1 × load buffer*.
5. Dilute the RNA sample to 250 μl with H_2O and leave at 65°C for 5 min.
6. Add 250 μl *2 × load buffer* and apply the sample to the column.
7. Collect the flow-through, leave it at 65°C for 5 min and reapply to the column.

8. Wash the column with 5 ml *1 × load buffer.*[b]
9. Wash the column with 2 ml *medium-salt buffer.*[c]
10. Elute poly-A^+ RNA with 2 ml *elution buffer*[d] and place on ice.
11. Extract with phenol-chloroform and ethanol-precipitate.
12. Redissolve in 50 μl double-distilled H_2O and store at −80°C.
13. After use, regenerate the oligo(dT)-cellulose resin with 10 ml *wash buffer*, 10 ml H_2O and 10 ml *1 × load buffer*. Store the column at 4°C.

Notes

a. 0.2 g oligo(dT)-cellulose powder gives rise to about 1 ml bed-volume with a capacity of about 500 μg polyadenylate. Flow-rates can be improved by suspending in a larger volume and removing fines prior to pouring the column, but we find this step unnecessary.
b. The wash contains poly-$A^−$ RNA and should be collected at least in a new series of experiments.
c. The medium-salt wash improves the purity of poly-A^+ RNA and is a potential source of mRNAs with oligo-A tails.
d. The majority of poly-A^+ RNA is eluted with 1 ml elution buffer.

Comments

The amount of purified poly-A^+ RNA is often so small that quantification by A_{260} measurements in a standard spectrophotometer is too wasteful. Instead the purified poly-A^+ RNA can be coelectrophoresed with total RNA and subjected to Northern analysis with a probe for an abundant mRNA such as eEF-1α or GAPDH. Moreover, the Northern analysis has the added advantage of examining the integrity of the purified poly-A^+ RNA.

2.3. In vitro synthesis of RNA

In vitro transcription provides DNA-directed synthesis of μg to mg amounts of RNA in a single reaction. Numerous vectors have been developed for *in vitro* transcription. The majority of these vectors contain the promoter regions for the bacteriophage RNA polymerases T3, T7 or SP6 adjacent to a polylinker for insertion of the desired fragments. In order to transcribe the insert alone, DNA must be linearised by cleaving downstream of the insert. As an alternative to liniarised plasmids one can use synthetic oligonucleotides containing the T7 promoter.[7] It is sufficient that the T7 promoter region is double-stranded to obtain efficient transcription (Fig. 2.1).

For optimal transcriptional efficiency of T7 polymerase a few points should be considered. The initial base in the transcript must be a purine and preferably a G (the optimal sequence is (5'-Gp(G/C)).[8] However, the polymerases accept a wide range of modifications at the 5' positions such as GpppG, dinucleotides and biotinylated dinucleotides (provided the 3'-nucleotide is a G, Section 2.3.2). If transcriptions are performed under conditions where one of the nucleotides is limiting the sequence of the first ten nucleotides is crucial for a high yield. If [α-^{32}P] UTP is limiting in cotranscriptional labelling of RNA uridines within the initial ten nucleotides of the transcript will result in a high frequency of short abortive transcripts. This has been rationalised as a result of a transition from a labile initiation complex (the initial ten nucleo-

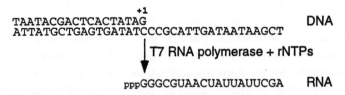

Fig. 2.1. Diagram illustrating an oligonucleotide-based template for transcription.

ides) to a very stable elongation complex distal of the initial ten nucleotides.

The basic transcription procedure including some of the optimisations introduced by Gurevich et al.[9] is described in Section 2.3.1, while alterations to the basic procedure involving incorporation of the cap-structure, or random incorporation of chemically derivatised nucleotides, or radioactive labelled nucleotides, are described in Section 2.3.2.

2.3.1. In vitro transcription of RNA

Materials
10 × polymerase buffer[a]
   100 mM NaCl
   80 mM $MgCl_2$
   20 mM spermidine
   800 mM Tris-HCl pH 8.0
T 7, T 3 or SP6 RNA polymerase
DTT 100 mM
Linearised DNA template[b]
10 × rNTP mix[c] (Pharmacia)
   10 mM of each rNTP

Preparative synthesis of RNA (500 μl reaction)[d]
1. Mix at room temperature in the following order:[e]
    20 μl   1 mg/ml *linearised DNA* (~10 nM final concentration; use 100 nM final concentration if template is composed of synthetic oligonucleotides)
    326 μl H_2O
    50 μl   *10 × polymerase buffer*
    50 μl   *100 mM DTT*
    50 μl   *10 × rNTP mix*[f,g]
    4 μl   *T 7 RNA polymerase* (20 units/μl)
2. Mix and incubate at 37°C for 1–2 h.[h]
3. Extract 2× with phenol.

4. Extract 2× with chloroform.
5. Add 50 µl 3.0 M NaOAc, pH 6.0.[i]
6. Add 2.5 vol EtOH (−20°C).
7. Precipitate and wash in 70% EtOH.
8. Redissolve in H_2O.

Notes

a. T3 and T7 polymerase activities do not require NaCl.
b. The *in vitro* transcribed RNA molecules usually have heterogeneous 3'-ends. In order to reduce the heterogeneity the DNA should be linearised with a restriction enzyme which generates a 5'-overhang. The 3'-overhangs generate frayed ends and promote snap-backs resulting in transcription of the sense strand. The 3'-overhang can be removed by treatment with the Klenow polymerase in the presence of 0.25 mM dNTPs. It is important to phenol extract the DNA after the restriction digest to remove traces of RNases present in some restriction enzyme preparations. As an alternative to linearised plasmid one can use synthetic oligonucleotides containing the T7 promoter.[7]
c. The nucleotides must be pH-equilibrated (pH 7.5) before use.
d. Trace amounts of $[\alpha\text{-}^{32}P]$ NTP, $[\alpha\text{-}^{35}S]$ NTP or $[\alpha\text{-}^{14}C]$ NTP may be included in the preparative RNA synthesis in order to quantify the final yield.
e. The DNA template may precipitate if 10 × polymerase buffer is added directly.
f. rNTPs in the preparative reaction mix may be raised to 5 mM final concentration resulting in a higher yield. Adjust the Mg^{2+} concentration accordingly.[g]
g. The free $[Mg^{2+}]$ must be adjusted depending on the nucleotide concentration. Since each nucleotide chelates one Mg^{2+} ion a rule of thumb is: The $[Mg^{2+}]$ should exceed the total nucleotide concentration by approximately 4 mM.
h. To eliminate contaminating DNA, 2 µl RNase-free DNase can be added and a further incubation of 10–30 min be carried out.
i. Alternatively, to reduce the precipitation of the nucleotides, 0.2

vol 10 M ammonium acetate and 2 vol EtOH at RT can be used for the precipitation.

Comments

The *in vitro* synthesised RNA can be purified in a denaturing polyacrylamide gel or by gel filtration (spin column, Section 2.1.2). In the latter case, the DNA template is not removed and a DNase step may be required for some applications.

2.3.1.1. Quantification of in vitro transcribed RNA (in the presence of a trace amount of $[\alpha\text{-}^{32}P]$ UTP)

Materials

Whatman DE-81 filter discs
Scintillation vial
0.5 M Na_2HPO_4 solution (nontitrated)

Procedure

1. Mix 2 μl of the reaction mixture (after step 2 in Section 2.3.1) with 18 μl H_2O.
2. Spot 8 μl to each of two *Whatman DE-81* filter discs. Air dry.
3. Place one filter in a *scintillation vial* and count (count A).
4. Place the second filter in 20–25 ml of *0.5 M Na_2HPO_4 solution (nontitrated)*. Swirl gently for 2–3 min. Pour off buffer.
5. Repeat wash 2–3 times.
6. Wash twice in H_2O for 30 s each time.
7. Rinse in EtOH, dry and count (count B).

Assuming equal representation of the different nucleotides. RNA synthesised (in μg) = 1.24 × (count B/count A) × (nmols cold UTP in the reaction).

2.3.2. Cotranscriptional labelling or modification of RNA

Radioactive or modified nucleotides can be incorporated randomly in a transcript or specifically at the 5′-position of the transcript during transcription. Thus it is feasible to incorporate cap-structures

(GpppG), dinucleotides (NpG), terminal monophosphate (pG, thio-G) and photoreactive nucleotides at the 5'-end. However, incorporation of nucleotides with ribose modifications requires some adjustments as described in Section 2.3.2.2.

Materials
10 × polymerase buffer:[a]
  100 mM NaCl
  80 mM $MgCl_2$
  20 mM spermidine
  800 mM Tris-HCl, pH 8.0
10 × labelling rNTP mix ([α-^{32}P] UTP)
  5 mM ATP
  5 mM CTP
  5 mM GTP
  0.5 mM UTP (or 50 μM UTP if high specificity is required)
[α-^{32}P] UTP (⩾700Ci/mmol)
10 × rNTPs for biotinylation
  5 mM NTP
0.5 mM *Biotin 16-UTP* (Boehringer–Mannheim)
10 × GpppG capped rNTP mix
  10 mM GpppG (or GpppG–2'-methyl) (Pharmacia)
  5 mM ATP
  5 mM CTP
  0.5 mM GTP
  5 mM UTP
DTT 100 mM

Procedure
1. Mix in the following order:
    1 μl 1 mg/ml *linearised DNA* (~10 nM final concentration. Use 100 nM final concentration if template is composed of synthetic oligonucleotides).
    7.5 μl H_2O
    2 μl *10 × polymerase buffer*

2 µl *100 mM DTT*
2. Add **either**
   For randomly-labelled RNA synthesis[b]
   2 µl   *10 × rNTPs labelling mix*
   5 µl   *[α-^{32}P] UTP* (≥700 Ci/mmol)
   or for biotinylated RNA transcripts.[c]
   2 µl   *10 × rNTPs for biotinylation.*
   2 µl   0.5 mM *biotin 16-UTP.*
   3 µl H_2O.
   or for cap-modified RNA transcripts.[d]
   2 µl   *10 × GpppG capped rNTP mix.*
   5 µl H_2O.
3. Add 0.5 µl *T 7 RNA polymerase* (20 units/µl).
4. Mix and incubate at 37°C for 1 h.
5. Add 180 µl 0.25 M NaOAc (pH 6.0).
6. Add 1 µl 5 mg/ml tRNA (*E. coli* RNase free) or glycogen (5 mg/ml) if tRNA is critical.
7. Extract 1 × with phenol.
8. Extract 1 × with chloroform.
9. Add 500 µl EtOH and precipitate.
10. Wash with 70% EtOH.
11. Redissolve in H_2O.

Notes
a. Consult the Notes in Section 2.3.1.
b. RNA labelled to a high specific activity is unstable and should be used within a day if full length RNA is required.
c. Similar procedures can be used for other modified nucleotides. The variations in the affinity of the modified nucleotide to the RNA polymerase usually require adjustments for each experiment (see Section 2.3.2.2).
d. The ratio of cap analogue to GTP determines the ratio of capped transcripts at the 5'-end of the transcript.[10] Excess of the cap analogue does not affect the elongation of the transcript al-

though the reduction in GTP concentration will reduce the overall efficiency of the transcription.

Comments
The transcription efficiency is greatly reduced or abolished if the 5'-nucleotide is not a guanosine. Some transcripts require nucleotides different from pppG at the 5'-end, and in these cases special precautions should be taken. If the second nucleotide is a G, ribodinucleotides, such as ApG, ^{4S}UpG, etc. can be added to the reaction to cap the transcript. The ribodinucleotide will be incorporated at the 5'-end at the same ratio as between the dinucleotide and GTP in the reaction mixture. An alternative method of generating any 5'-end is given in Section 2.3.2.1.

Moreover, addition of diribonucleotides to the transcription mixture is convenient if the transcript at a later stage is going to be 5'-end labelled, thus avoiding the dephosphorylation step. The transcript can also be 5'-end labelled directly by the addition of $[\gamma\text{-}^{32}P]$ GTP to the reaction mixture.

2.3.2.1. Generation of a specific terminus by RNase H
A general method to generate precisely defined ends of the transcript utilises the RNase H-mediated cleavage of RNA in DNA:RNA hybrids. This approach can with advantage be used for transcripts aimed at internal ligations (Section 2.3.4)[11] or for production of high yields of transcripts lacking the optimal consensus sequence at the 5'-end of the transcript.[12]

RNA can be cleaved at a particular site by RNase H if the RNA transcript is hybridised to a chimeric RNA/DNA oligonucleotide composed of four consecutive deoxynucleotides flanked by 2'-O-methyl RNA on either side[13] or at the 3'-site (Fig. 2.2). The RNA will be cleaved at a site opposite to the 5'-end of the DNA segment. The location of the four base deoxy-segment at the 5'-end of the chimeric oligonucleotide can be used as a more general method to process *in vitro* transcribed RNA. When this approach is applied, the DNA template encoding the RNA to be studied can be cloned

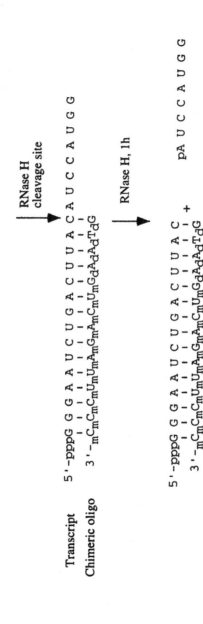

Fig. 2.2. Diagram of site-specific processing of the 5'-terminus by RNase H cleavage. The chimeric primer is made of 2'-O-methoxynucleotides and deoxynucleotides. The processed transcript contains a 5'-phosphate.

after a 15 base region optimal for T7 transcription.[8] RNase H cleavage in the presence of a chimeric oligonucleotide, complementary to the initial 15 nucleotides of the transcripts and with the DNA segment at the 5'-end, will generate an RNA without any sequence restrictions. The reaction can either be performed in solution or with an immobilised chimeric primer.[12]

Material
2 × RNase H buffer
    40 mM Tris-HCl, pH 7.5
    20 mM $MgCl_2$
    200 mM KCl
    20 mM DTT
    10% sucrose
RNase H (Boehringer Mannheim)
RNasin

Procedure
1. Mix in a total volume of 10 μl the RNA[a] and the chimeric RNA/DNA oligonucleotide at a ratio (1:1.5) in TE.
2. Denature at 95°C for 3 min.
3. Anneal for 10 min at 37°C.
4. Add 10 μl *2 × RNase H buffer*; 40 units *RNasin*; 5 units *RNase H*.
5. Incubate for 1 h at 37°C.
6. Add 120 μl TE.
7. Extract 1 × with phenol.
8. Extract 1 × with chloroform.
9. Precipitate with 1/10 vol 3 M NaOAc (pH 6.0) and 2.5 vol EtOH.

Notes
a. The reaction can accommodate upto 5 μg/μl RNA.

2.3.2.2. Cotranscriptional incorporation of nucleotides containing modified ribose moieties

Random incorporation of modified residues in *in vitro* transcribed RNA is of particular interest in analogue interference experiments (Section 4.6.3) where randomly modified RNA molecules with altered properties are selected and compared to the original pool of RNA molecules. A crucial prerequisite for the interference analysis is the ability to identify the modified residues. A convenient and efficient method is the use of nucleotides which contain both the modification of interest (e.g. 2'-O-methyl NTP, dNTP, inosinetriphosphates) and an α-phosphorothioate group (one of the non-bridging oxygens is replaced with a sulphur).[14-16] The concurrent incorporation of the phosphorothioate renders the modified nucleotides sensitive to a selective cleavage by iodine.

The ease of incorporation of the modified nucleotide is highly dependent on the type of modification. Base modifications are usually incorporated in the normal transcription buffers, whereas, changes in the ribose moiety require altered buffer conditions.

Phosphate modifications: T7 polymerase does not discriminate significantly between NTP and [αS] NTP (NEN). To obtain the appropriate modification level the NTP is substituted with [αS] NTP to the desired level. A standard transcription protocol, such as the one described in Section 2.3.1, can be used.

Concurrent base and phosphate modification: The current examples of cotranscriptional incorporation of modified bases are rather limited,[14-16] and the modified NTPs are generally not commercially available. The standard buffer conditions can be employed, but the modified nucleotides are incorporated with different efficiencies e.g. 5% incorporation of inosine at the G positions requires 40% inosinetriphosphates in the reaction mixture, while 7-methyl-guanosine is incorporated as readily as GTP.

Concurrent ribose and phosphate modifications: Modifications at the 2'-hydroxyl group (i.e. dNTP or 2'-O-methyl NTP) are very informative for the analysis of minor groove interactions and tertiary interaction. The incorporation requires special buffers as described below.

Materials

2 × modification transcription buffer
   80 mM Tris-HCl, pH 8.0
   4 mM spermidine
   20 mM DTT
   16% PEG 8000
   0.02% Triton X-100
   5 mM MnCl$_2$[a]

5 × nucleotide mix[b]
   5 mM of each of the three unmodified nucleotides
   A total of 5 mM of the 2'-modified nucleotide and the corresponding unmodified nucleotide

NTP, dNTP and [α-^{35}S] dNTP (Pharmacia)
NTPαS (Amersham or NEN)
2'-O-methyl-NTP and *2'-O-methyl-NTPαS* can be made as described by Gaur et al.[17]
Ammonium acetate, 4 M
tRNA (5 mg/ml)

Procedure

1. Mix in the following order:
   1 μl 1 mg/ml *linearised DNA* (~10 nM final concentration. Use 100 nM final concentration if template is composed of synthetic oligonucleotides (Section 2.3.1).
   4 μl H$_2$O.
   4 μl *nucleotide mix*.
   10 μl *2 × modification transcription buffer*.
   1 μl *T 7 RNA polymerase* (40 units/μl).
2. Incubate at 37°C for 1 h.
3. Add 80 μl H$_2$O.

4. Add 1 μl 5 mg/ml tRNA (*E. coli* RNase free) or glycogen (5 mg/ml) if tRNA is critical.
5. Extract 1× with phenol.[c]
6. Extract 1× with chloroform.
7. Add 100 μl 4 M ammonium acetate and 500 μl EtOH; precipitate and wash in 70% EtOH.[d]
8. Redissolve in H_2O.

Notes

a. Add 5 mM $MgCl_2$ (in addition to $MnCl_2$) for incorporation of 2′-modified CTP.
b. The fraction of modified NTP in the transcription mixture to obtain 5% incorporation should be approximately 50% for dCTP and dTTP, while more than 75% of the nucleotide should be modified for dATP, dGTP, [αS] dNTP or 2′-O-methyl NTP. For replacement of all the riboguanosines or ribocytidines with the 2′-derivative there should not occur consecutive stretches of these nucleotides in the initial ten bases of the transcript. For replacement of all riboadenosines, adenosines should not appear among the initial ten positions of the transcript.
c. The phenol and chloroform extraction can be omitted if the RNA is to be gel-purified
d. Precipitation with ammonium acetate reduces the coprecipitation of nucleotides.

2.3.3. End-labelling of RNA

2.3.3.1. Dephosphorylation of RNA
In vivo or *in vitro* synthesised RNA generally contains a 5′-phosphate(s) whereas RNA fragments generated by nuclease digestion usually contain a 3′-phosphate (Section 4.4). These phosphates must be removed prior to labelling of the respective ends. The same procedure may be used for dephosphorylations of either end.

Materials

10 × CIP buffer

  500 mM Tris-HCl, pH 7.5

  10 mM $MgCl_2$

  1 mM $ZnCl_2$

Calf Intestinal Phosphatase (RNase free 10 units/μl, Boehringer Mannheim)

EGTA 200 mM

Procedure

1. Dissolve 0.5–5 μg RNA in 10 μl H_2O.
2. Denature RNA at 95°C for 30 s, then on ice.
3. Add 40 μl phosphatase mix (5 μl *10 × CIP buffer*, 35 μl H_2O, 10 units *Calf Intestinal Phosphatase*).
4. Incubate at 37°C for 15 min followed by incubation at 56°C for 15 min.
5. Add an additional 10 units of *Calf Intestine Phosphatase*.
6. Repeat step 4.
7. Add 2 μl 200 mM EGTA.
8. Heat inactivate at 95°C for 4 mins—then on ice.
9. Add 100 μl H_2O and 15 μl 3.0 M NaOAc, pH 6.0.
10. Extract 2× with phenol.[a]
11. Extract 2× with chloroform.
12. Add 2.5 vol EtOH.
13. Precipitate and wash with 70% EtOH.

Notes

a. The heat inactivation does not completely inactivate the phosphatase.

Comments

The dephosphorylation of *in vitro* transcribed RNA can be avoided if a diribonucleotides, such as CpG, ApG or UpG is included in the transcription mixture. The diribonucleotides will be incorporated at the 5′-end of the transcript (see Section 2.3.2).

2.3.3.2. 5'-end labelling of RNA
Materials
10 × Polynucleotide kinase buffer
    500 mM Tris-HCl, pH 8.0
    100 mM MgCl₂
    50 ng/ml BSA
    50% glycerol
    100 mM DTT
T 4 Polynucleotide kinase
EDTA 0.5 M, pH 8.0
[γ-³²P] ATP (3000 Ci/mmol, 10 μCi/μl)

Procedure
1. Dissolve 0.5–5 μg dephosphorylated RNA in 10 μl H₂O.
2. Denature RNA at 95°C for 30 s, then on ice.
3. Add 8 μl [γ-³²P] ATP (3000 Ci/mmol, 10 μCi/μl) in H₂O.[a]
4. Add 2 μl 10× Polynucleotide kinase buffer and mix.
5. Add 1 μl T4 Polynucleotide kinase (5 U/μl).
6. Incubate at 37°C for 30 min, then on ice.
7. Add 1 μl 0.5 M EDTA.
8. Add 80 μl 0.3 M NaOAc, pH 6.0.[b]
9. Precipitate and wash with 70% EtOH.

Notes
a. Lyophilise [γ-³²P] ATP if dissolved in ethanol solution.
b. Precipitation is optional if gel filtration methods are used for the purification.

Comments
The end-labelled RNA can be purified in a denaturing polyacrylamide gel (*n* < 1000) or an agarose gel (*n* > 1000) or by gel filtration in Sephadex G–50. The phenol and chloroform extraction should be included after the gel filtration. If the RNA is used without purification a phenol and chloroform extraction should be included before the precipitation step.

2.3.3.3. 3'-end labelling of RNA

3'-end labelling of RNA can either be done by ligating $[\alpha\text{-}^{32}P]$ pCp to the 3'-end using RNA ligase,[18] or adding a 3'-deoxynucleotide such as $[\alpha\text{-}^{32}P]$cordycepin-5'-triphosphate using poly(A) polymerase. The advantage of the ligase method is that radioactive pCp is relatively easy to make from standard radioactive nucleotides.

2.3.3.3.1. RNA ligase method.

Materials

$2 \times$ RNA ligase buffer
  100 mM HEPES-KOH, pH 7.5
  6.6 mM DTT
  30 mM $MgCl_2$
  20% DMSO
  0.02 mg/ml BSA
$[\alpha\text{-}^{32}P]$ pCp (3000 Ci/mmol, Amersham or Section 2.3.3.4)
T4 RNA ligase (5 U/µl)
ATP, 10 mM

Procedure

1. Dilute 5–20 pmol RNA in water. Adjust H_2O so the total volume of the ligation will be 20 µl.
2. Denature RNA at 95°C for 30 s and cool on ice.
3. Mix the following in order:
   20 µCi $[\alpha\text{-}^{32}P]$ pCp (3000 Ci/mmol) or 5 µl of pCp mix (Section 2.3.3.4)
   10 µl 2 × RNA ligase buffer
   2 µl ATP (10 mM)
   1 µl T4 RNA ligase (5 U/µl)
4. Leave overnight at 0°C or 2 h at RT.
5. Add load buffer.

2.3.3.3.2. Poly (A) polymerase method

Materials

$10 \times$ Poly A polymerase solution

670 mM $MgSO_4$

20 mM $MnCl_2$

[α-^{32}P] *cordycepin-5'-triphosphate*

Poly A polymerase (300–600 units/μl)

Procedure

1. Dissolve 100 pmol RNA in 8 μl 50 mM Tris-HCl, pH 7.5.
2. Denature at 95°C for 30 s and cool on ice.
3. Dissolve 100 μCi lyophilised [α-^{32}P] *cordycepin-5'-triphosphate* with the denatured RNA solution.
4. Add 1 μl *10 × Poly A polymerase solution.*
5. Add 1 μl *Poly A polymerase.*
6. Incubate at 37°C for 30 min.
7. Terminate the reaction by addition of 50 μl 0.3 M NaOAc (pH 6.0) followed by 150 μl EtOH (leave on dry ice for 15 min) for precipitation.
8. Dissolve the pellet in loading buffer.[a]

Notes

a. The end-labelled RNA can be purified in a denaturing poly-acrylamide gel ($n < 1000$) or an agarose gel ($n > 1000$) or by gel filtration in Sephadex G–50. A phenol and chloroform extraction should be included after the gel filtration. If the RNA is used without purification a phenol and chloroform extraction should be included before the precipitation. Precipitation is op-tional if gel filtration methods are used for the purification.

2.3.3.4. Preparation of [α-^{32}P] pCp from Cp and [γ-^{32}P] ATP

Materials

10 × Polynucleotide kinase buffer

    500 mM Tris-HCl, pH 8.0

    100 mM $MgCl_2$

    0.005% BSA

    50% glycerol

    100 mM DTT

[γ-^{32}P] *ATP* ($\geq$3000 Ci/mmol)
ATP, 25 μM
Cp, 3 mM (Sigma)
T 4 Polynucleotide kinase (5 units/μl)

Procedure
Mix:
   100 μCi [γ-^{32}P] *ATP* (3000 Ci/mmol).
   1 μl 25 μM ATP (can be omitted if higher specific activity is needed).
   2 μl 10 $\times$ *polynucleotide kinase buffer*.
   1 μl 3 mM *Cp*.
   H$_2$O to 20 μl.
   1 μl *T 4 Polynucleotide kinase* (5 U/μl).
Incubate at 37°C for 30 min and heat-inactivate at 70°C for 5 min.

Comment
The [α-^{32}P]pCp need not be purified before use in the ligase reaction.

2.3.4. Specific modification of RNA at an internal site

In many structural and functional studies it is desirable to modify a specific site of an RNA. Site-specific modifications at internal positions of small RNAs (<20 nucleotides) is most conveniently obtained by introducing a modified nucleotide during chemical synthesis by standard phosphoramidite chemistry and will not be covered in this book. A wide spectrum of modified ribo- and deoxynucleotide analogues are commercially available for chemical synthesis. Moreover, custom-made RNA oligonucleotides containing O-methyl, O-allyl, or O-silyl protection of the 2'-hydroxyl group of the ribose can be obtained from commercial sources. For crosslinking purposes, 5-bromo-U- or 5-iodo-U phosphoramidites are available. However, depending on the type of modification, the 5'–3'-coupling efficiency is typically decreased to 90–97% which

A.

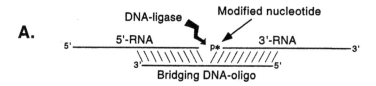

B.

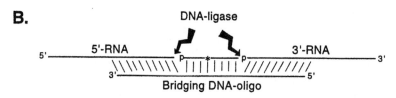

Fig. 2.3. The principle behind site-specific modification of RNA. A, Ligation of two RNAs. In the illustrated example, the 3'-RNA contains a 5'-phosphorylated modified nucleotide at the 5'-end. B, Ligation of three RNAs. The short centrally positioned RNA, which can be synthesised chemically, contains the modified nucleotide. In both examples, the RNAs are aligned by using a bridging DNA oligonucleotide, and the ligation(s) is catalysed by T4 DNA ligase.

means that only a fraction of RNAs larger than 15–20 mers is chemically homogeneous. This method is therefore not useful for large RNAs.

Site-specific modification of large RNAs (>20 nucleotides) has until recently been a difficult task which involved an RNA ligase-mediated ligation of two RNAs. An alternative method has been developed by Moore and Sharp[19] which has facilitated this reaction.

The basic principle is outlined in Fig. 2.3A: Two RNAs, a 5'-RNA and a 3'-RNA, containing the sequences 5' and 3' of the site to be modified, respectively, are synthesised so that the 3'-RNA contains the 5'-phosphorylated modified nucleotide at the 5'-end. A complementary DNA oligonucleotide is annealed to the two RNA halves, and the ligation step is catalysed by T4 DNA ligase, which will join the 3'-OH of the 5'-RNA to the 5'-terminal mono-phosphate of the 3'-RNA. The 5'-RNA may be synthesised by standard *in vitro* T7 RNA polymerase run-off transcription which will produce a 3'-OH terminal (Section 2.3.1). If the 3'-RNA is synthesised by *in vitro* transcription action must be taken to replace the triphosphate with monophosphate at the 5'-terminus. One option is to dephosphorylate and rephosphorylate the 3'-RNA (Section 2.3.3). This approach will allow incorporation of radioactive ^{32}P at the junction between the 5'- and 3'-RNA (see procedure below). Alternatively, the 3'-RNA may be initiated with a GMP during *in vitro* transcription or with a dinucleotide (NpG) and then phosphorylated (see Section 2.3.2). Cotranscriptional inclusion of a modified 5'-terminal residue will allow the incorporation of a large variety of modifications, including photoreactive crosslinkers and phosphorothioates, in the 3'-RNA, immediately downstream from the site of ligation (see Section 2.3.2). The 5'- and 3'-RNAs may also be modified individually throughout the transcript by including radioactive or chemically modified NTPs during the transcription process. A limitation of the approach is the requirement for a G-residue at the site of transcription initiation, in order to obtain sufficient RNA product. A solution to this problem is to use RNase H to cleave off the 5'-terminal region of the RNA as described in Section 2.3.2.1.

The most flexible approach is to perform a double ligation of three RNAs placing a synthetic RNA oligonucleotide in the middle[a] (Fig. 2.3B).[20] Although this reaction is slightly more inefficient it allows incorporation of a greater variety of modified nucleotides at any position of the centrally positioned RNA oligonucleotide and eliminates the requirement of having G-residues immediately downstream from the site of modification.

A general problem when ligating RNAs using RNA ligase is the heterogeneous 3'-ends of *in vitro* synthesised RNAs, which give rise to a variable number of nucleotides at the site of ligation. This problem is circumvented in the Moore and Sharp method since only RNAs which match the complementary DNA oligonucleotide perfectly is a substrate for T4 DNA ligase. A frayed end of the 5'-

RNA will therefore lower the yield of ligated product but it will not influence the homogeneity of the product.

In the Moore and Sharp[19] procedure outlined below a radioactive phosphate is inserted at the junction by labelling the 3'-RNA using T4 polynucleotide kinase and $[\gamma\text{-}^{32}P]$ATP, but the ligation approach is useful for many other types of modifications. Other applications include:

(1) Random labelling of sections of the RNA (patch labelling).[19]
(2) Insertion of 2'-ribose- and base-modified nucleotides and/or phosphorothiorates nucleotides for functional studies.[19,21,22]
(3) Insertion of a radioactive phosphate at a specific site for label-transfer by UV crosslinking.[23]
(4) Inserting a photoreactive nucleotide for long wavelength UV crosslinking (see also Section 4.3.2).[24]
(5) Circularisation of RNA.[25]

Materials

5'- and 3'-RNAs to be ligated (20 μM stocks in H_2O)
DNA bridging oligonucleotide (20 μM stock in H_2O)
10 × Ligase buffer
  500 mM Tris-HCl, pH 7.5
  100 mM $MgCl_2$
  200 mM DTT
  0.5 mg/ml BSA
T4 Polynucleotide kinase (10 units/μl)
T4 DNA ligase (10 USB units/μl, note that the unit definition varies among different suppliers)
$[\gamma\text{-}^{32}P]$ *ATP* (10 mCi/ml, 3000 Ci/mmol = 3.33 μM)
ATP 10 mM
DNAase (RNase-free)
Formamide loading buffer:
  80% formamide
  0.1% xylene cyanol

0.1% bromophenol blue
1 mM EDTA, pH 8.0.

Procedure

1. Dry down 5 μl of $[\gamma\text{-}^{32}P]$ *ATP* (50 μCi, ~10 pmol) in a sil-iconised 1.5 ml Eppendorf tube.
2. Add to the tube in the following order:
   0.5 μl *10 × ligase buffer*
   3 μl *3'-RNA* (60 pmol dephosphorylated; see Section 2.3.3.1)[b]
   1.0 μl H_2O
   0.5 μl *T4 Polynucleotide kinase*
   (final volume: 5.0 μl)
3. Incubate for 60 min at 37°C.
4. Heat inactivate the kinase by incubating for 10 min at 75°C and store on ice.
5. Add:
   1.0 μl *10 × Ligase buffer*
   3.0 μl *5'-RNA* (60 pmol)[b]
   3.0 μl 20 μM *DNA bridging oligonucleotide* (60 pmol)
   (total volume: 12 μl)
6. Hybridise the two RNA halves to the DNA by incubating the reaction for 2 min at 75°C, followed by 5 min at 30°C.
7. Add:
   1.0 μl 10 mM *ATP*
   2.0 μl T4 *DNA ligase*[c]
   (total volume: 15 μl)
8. Incubate for 2–4 h at 30°C to ligate RNAs.
9. Add 1 μl of *DNase (RNase-free)* and continue to incubate for 15 min at 37°C.
10. Add 15 μl *formamide loading buffer* and load on a 0.5–1-mm thick denaturing polyacrylamide gel. It is convenient to load radioactively labelled full-length RNA, corresponding to the ligated product as a marker.

Notes

a. When ligating three RNAs the concentration of the RNAs may have to be optimised individually. A good starting point is to add equimolar amounts of each RNA.

b. If high specific activity is required, the kinase reaction should be carried out under conditions in which the 3'-RNA is in 2–3-fold molar excess over the labelled ATP as outlined here. The unlabelled RNA which lacks a 5'-phosphate will not participate in the subsequent ligation process. If the molar quantity of 3'-RNA is critical the molar ratio of ATP and 3'-RNA should be reversed by adding cold ATP such that the concentration is more than 4-fold higher than the concentration of 3'-RNA. Alternatively, if no radioactive label is required at the ligation junction, the 3'-RNA can be synthesised with a 5'-terminal phosphate or transcribed with GMP (Section 2.3.2).

c. T4 DNA ligase is not turned over efficiently in the RNA ligation process, so it must be added in stoichiometric amounts to the RNA.

Comments

A modified version of the Moore and Sharp method, introducing a photoreactive 4-thiouridine into specific positions in a large RNA molecule, was recently published by Yu and Steitz.[11] In this protocol the ^{32}P-$p^{4S}Up$ is ligated to the 3'-end of the 5'-half RNA using RNA ligase prior to ligation to the 3'-half RNA by DNA ligase and bridging oligonucleotide.

References

1. Chirgwin, J.M., Przybyla, A.E., MacDonald, R.J. and Rutter, W.J. (1979). Isolation of biologically active ribonucleic acid from sources enriched in ribonuclease. Biochemistry 18, 5294–5299.
2. Chomczynski, P. and Sacchi, N. (1987). Single-step method of RNA iso-

lation by acid guanidinium thiocyanate-phenol-chloroform extraction. Anal. Biochem. *162*, 156–159.

3. Kohrer, K. and Domdey, H. (1991). Preparation of high molecular weight RNA. Meth. Enzymol. *194*, 398–405.

4. Berger, S.L. (1987). Isolation of cytoplasmic RNA: ribonucleoside-vanadyl complexes. Meth. Enzymol. *152*, 227–234.

5. Aviv, H. and Leder, P. (1972). Purification of biologically active globin messenger RNA by chromatography on oligothymidylic acid-cellulose. Proc. Natl. Acad. Sci. USA *69*, 1408–1412.

6. Vauti, F. and Siess, W. (1993). Simple method of RNA isolation from human leucocytic cell lines. Nucleic Acid. Res. *21*, 4852–4853.

7. Milligan, J.F., Groebe, D.R., Witherell, G.W. and Uhlenbeck, O.C. (1987). Oligoribonucleotide synthesis using T7 RNA polymerase and synthetic DNA template. Nucleic Acids Res. *15*, 8783–8798.

8. Milligan, J.F. and Uhlenbeck, O.C. (1989). Synthesis of small RNAs using T7 polymerase. Meth. Enzymol. *180*, 51–62.

9. Gurevich, V.V., Pokrovskaya, I.D., Obukhova, T.A. and Zozulya, S.A. (1991). Preparatine *in vitro* mRNA synthesis using SP6 and T7 polymerases. Anal. Biochem. *195*, 207–213.

10. Konarska, M.M., Padgett, R.A. and Sharp, P.A. (1984). Recognition of cap structure in splicing *in vitro* of mRNA precursors. Cell *38*, 731–736.

11. Yu, Y. and Steitz, J.A. (1997). A new strategy for introducing photoactivatable 4-thiouridine (^{4S}U) into specific positions in a long RNA molecule. RNA *3*, 807–810.

12. Lapham, J. and Crothers, D.M. (1996). RNaseH cleavage for processing of *in vitro* transcribed RNA for NMR studies and RNA ligation. RNA *2*, 289–293.

13. Inoue, H., Hayase, Y., Iwai, S. and Ohtsuka, E. (1987). Sequence dependent hydrolysis of RNA using modified oligonucleotide splints and RNase H. FEBS Lett. *215*, 327–330.

14. Strobel, S.A. and Shetty, K. (1997). Defining the chemical groups essential for Tetrahymena group I intron function by nucleotide analog interference mapping. Proc. Natl. Acad. Sci. USA *94*, 2903–2908.

15. Conrad, F., Hanne, A., Gaur, R.K. and Krupp, G. (1995). Enzymatic synthesis of 2′-modified nucleic acids: identification of important phosphate and ribose moieties in RNase P substrates. Nucleic Acids Res. *23*, 1845–1853.

16. Kahle, D., Wehmayer, U., Char, S. and Krupp, G. (1990). The methylation of one specific guanosine in a pre-tRNA prevents cleavage by RNase P and by the catalytic M1 RNA. Nucleic Acids Res. *18*, 837–844.

17. Gaur, R.K., Sproat, B.S. and Krupp, G. (1992). Tetrahedron Lett. *33*, 3301–3304.

18. Bruce, A.G. and Uhlenbeck, O.C. (1978). Reactions at the termini af tRNA with T4 RNA ligase. Nucleic Acids Res. 5, 3665–3677.
19. Moore, M.J. and Sharp, P.A. (1992) Site-specific modification of pre-mRNA: the 2′-hydroxyl groups at the splice sites. Science 256, 992–997.
20. Query, C.C., Moore, M.J. and Sharp, P.A. (1994). Branch nucleophile selection in pre-mRNA splicing: evidence for the bulged duplex model. Genes Dev. 8, 587–597 (1994).
21. Moore, M.J. and Sharp, P.A. (1993). Evidence for two active sites in the spliceosome provided by stereochemistry of pre-mRNA splicing. Nature 365, 364–368.
22. Query, C.C., Strobel, S.A. and Sharp, P.A. (1996). Three recognition events at the branch-site adenine. EMBO. J 15, 1392–1402.
23. Gozani, O., Feld, R. and Reed, R. (1996). Evidence that sequence-independent binding of highly conserved U2 snRNP proteins upstream of the branch site is required for assembly of spliceosomal complex A. Genes Dev. 10, 233–243.
24. Wyatt, J.R., Sontheimer, E.J. and Steitz, J.A. (1992). Site-specific cross-linking of mammalian U5 snRNP to the 5′-splice site before the first step of pre-mRNA splicing. Genes Dev. 6, 2542–2553.
25. Chen, C.Y. and Sarnow, P. (1995). Initiation of protein synthesis by the eukaryotic translational apparatus on circular RNAs. Science 268, 415–417.

Preparation of Protein

3.1. Isolation of protein from cells

3.1.1. Preparation of HeLa cell extract

This procedure describes the preparation of nuclear and cytoplasmic extracts from cells grown in suspension (2–20 l) and is essentially as described by Dignam et al.[1] and modified by Jamison and Garcia-Blanco.[2] A procedure for small-scale preparation of extracts from HeLa cells grown as monolayer has been described by Lee and Green.[3] The latter procedure is recommended when the cell material is scarce, if radioactively labelled extracts are made, if several extracts from parallel cultures, or if expensive growth conditions are necessary for cell propagation.

Both methods involve swelling of the cells in hypotonic solution, disruption of the plasma membrane, and extraction of the nuclei in a high salt buffer. The extracts support a number of nuclear functions and have been extremely useful in studying transcription and splicing (see Section 5.1). Nuclear extracts are also a useful starting point for purification of snRNAs, snRNPs, hnRNP proteins and SR-proteins (Section 3.1.2).

Cells, nuclei, or nuclear extracts are also commercially available from various sources. Material from The National Cell Culture Center, 8500 Evergreen Boulevard, Minneapolis, MN 55433 USA and Computer Cell Culture Center sa, Place du Parc, 20 B–7000 Mons, Belgium, is recommended.

57

Materials

HeLa S3 spinner cells[a]

   8–16 l

PBS

   20 mM potassium phosphate, pH 7.4

   0.13 M NaCl

PMSF[b]

   Stock solution (0.13 g PMSF in 7.5 ml propanol)

DTT, 1 M

Buffer A:	100 ml (for 8 l prep)
10 mM HEPES-KOH, pH 7.9	1 ml 1 M
10 mM KCl	0.33 ml 3 M
1.5 mM MgCl$_2$	150 μl 1 M
0.5 mM DTT	50 μl 1 M
Buffer B:	10 ml (for 8 l prep)
0.3 M HEPES-KOH, pH 7.9	3 ml 1 M
1.4 M KCl	4.8 ml 3 M
30 mM MgCl$_2$	300 μl 1 M
0.5 mM DTT	50 μl 1 M
Buffer C:	10 ml (for 8 l prep)
20 mM HEPES-KOH, pH 7.9	0.2 ml 1 M
0.42 M NaCl	0.84 ml 5 M
1.5 mM MgCl$_2$	15 μl 1 M
0.2 mM Na$_2$EDTA	4 μl 0.5 M
0.5 mM DTT	5 μl 1 M
25% glycerol	4.1 ml 60% v/v
0.5 mM PMSF[b]	20 μl 0.25 M
Buffer D:	3 l (for 8 l prep)
20 mM HEPES-KOH, pH 7.9	60 ml 1 M
100 mM KCl	100 ml 3 M
0.2 mM EDTA	1.2 ml 0.5 M
1 mM DTT	1.5 ml 1 M
10% glycerol	500 ml 60% v/v
0.5 mM PMSF[b]	5 ml 0.25 M

Equipment

500-ml bottles and *Beckman JA10 rotor* (or equivalent)
50-ml disposable screw-cap tubes and *swinging bucket rotor*
15-ml tubes, 50-ml polycarbonate tubes and *Beckman JA20 rotor*
(or equivalent)
Dounce chamber and *pestles* (A and B) (Wheaton)
SW28 rotor and *polyallomer tubes* (or equivalent) if preparing S100
extract
Dialysis clips and *dialysis bag* (6000–8000 MW cutoff, autoclaved
and stored in 1 mM EDTA at 4°C).

Procedure

1. Prepare in advance: Buffers A, C, B (if making S100 extract)
   and D, leaving out the DTT. Store in a cold room. Add DTT
   to all buffers just before use. Autoclave or bake Dounce chamber and pestles. Store in a cold room. Precool all rotors.
2. Grow 8–16 l of *HeLa S3 spinner cells*.[a]
3. Harvest cells at a density of $5–8 \times 10^5 \, ml^{-1}$.
4. Count the number of cells to determine the amount of buffer
   C to be used later.
5. Centrifuge cells in *500 ml bottles* for 10 min at 1500 rpm using
   Beckman JA10 rotor or equivalent. **N.B.** Keep cells on ice
   during all steps and work in a cold room for the subsequent
   steps.
6. Pour off the supernatant and transfer pellets to 50 *ml polypropylene tubes*.
7. Centrifuge 10 min at 1500 rpm in a *swinging bucket rotor* at
   4°C.
8. Decant the remaining supernatant and mark cell pellet volume
   on the side of each tube. Be accurate, this is a critical step.
9. Wash the cells by resuspending in 5 volumes of the cell pellet
   volume (step 8) of cold *PBS*. Use pipettes to break up all
   clumps of cells but do *not* vortex or shake vigorously.

10. Centrifuge for 10 min at 1500 rpm in a *swinging bucket rotor* at 4°C and decant supernatant.
11. Resuspend cells in 5 volumes of the original pellet size of cold *buffer A* and let the cells swell for 10 min on ice.
12. Centrifuge for 10 min at 2500 rpm in a *swinging bucket rotor*.
13. Remove most of supernatant carefully with a 25-ml pipette. Cell volume should be approximately double in size.
14. Resuspend the swollen cell pellets in 2 volumes of the original cell pellet volume (step 8) of *buffer A*. Do not use pipettes to resuspend cells, but just loosen them enough so they can be transferred into a Dounce chamber for lysis.
15. Lyse cells in *Dounce chamber* using a tight pestle (*pestle A*). Make 10 strokes at medium speed, holding chamber as vertical as possible.[c] Rinse Dounce chamber with sterile deionised water before reuse in step 22.
16. Pour lysed cells into *50 ml polycarbonate tubes* for the *Beckman JA20 rotor* or equivalent.
17. Centrifuge for 15 min at 3000 rpm at 4°C.
18. Carefully remove supernatant with a pipette and save it for pre-S100 extract if desired (steps 28–32). The pellet will be soft and faint at this step.
19. Centrifuge pellets again in the same tubes for 20 min at 14,000 rpm at 4°C.
20. Remove supernatant and combine with the pre-S100 extract. The remaining pellet constitutes the crude nuclei.
21. Resuspend the pellet in 3 ml of *buffer C* per 10^9 cells, estimated from the cell count at the beginning (step 4). Release pellet from the bottom of the tube by swirling, the pellet will be broken up in the following step.
22. Transfer crude nuclei to Dounce chamber and apply ten strokes (medium speed) with a loose pestle (*pestle B*).
23. Pour lysed suspension into *15 ml polycarbonate tubes* and rotate at 4°C for 30 min. The suspension should be thick or stringy in appearance.

24. Centrifuge in *Beckman JA20* or equivalent for 30 min at 14,000 rpm. The cloudy supernatant is the nuclear extract.
25. Place the nuclear extract in a *dialysis bag*. Dialyse at 4°C against precooled *buffer D* as follows: 1 l for 1 h, fresh 1 l for 2 h, and finally fresh 1 l for 2 h.
26. Pour extract into a 50 ml tube and aliquot as desired. Small aliquots for testing and larger for storage. Protein concentration should be ~10 mg/ml.
27. Freeze immediately on dry ice, leave for at least 5 min, and store at −80°C.

Cytoplasmic S100 extract preparation:
28. Mix both pre-S100 extracts (steps 18 and 20).
29. Add 0.11 of the original cell volume (step 8) of *buffer B* and mix.
30. Centrifuge in *polyallomer tubes* at 130,000 g (r_{max}) for 60 min in an ultracentrifuge (27,000 rpm in a *Beckman SW28 rotor*).
31. Remove supernatant and add *DTT* to 1 mM final concentration.
32. Aliquot and store at −80°C.

Notes
a. The cells should be grown in 10% new-born calf or horse serum. It is very important that the cells are thriving with a low proportion of dead cells. Doubling times should be around 24 h.
b. For HeLa cells PMSF can be omitted without any significant effect on splicing efficiency but it is important for other cell types, e.g. B-cells.
c. Use a microscope to check that more than 80% of the cells are lysed. If lysis is unsatisfactory apply 10–20 more strokes and re-analyse the cells.

3.1.2. Purification of SR proteins from nuclear extract

SR proteins constitute a group of splicing factors that plays a major role in splice site recognition and assembly of the spliceosome. This group of proteins is characterised by common structural features, including a long stretch of serine-arginine repeats at the C-terminus, 1–2 RNA binding domains at the N-terminus and a phosphoepitope recognised by the monoclonal antibody mAB 104. They also share the functional capacity of complementing splicing-deficient S100 extracts in splicing (Section 3.1.1; see Fu et al.[4] and Valcárcel and Green[5] for a review). The metazoan SR-proteins fall into groups having conserved molecular weights of 20, 30 (three members), 40, 55 and 75 kDa. A rapid two step procedure has been published to purify SR proteins.[6] It utilises the characteristics that SR proteins are soluble in 65% ammonium sulphate and precipitated in 20 mM $MgCl_2$. The purification scheme below is similar to the Zahler et al. procedure[6] but modified so a nuclear extract can be used as starting material (see Section 3.1.1 for preparation of HeLa cell nuclear extract). Animal organs such as calf thymus or bovine brain can also be used as a cheaper starting material but the yields of individual SR-proteins are often highly variable. An alternative procedure which appears to yield additional SR-proteins has been published by Crispino and Sharp.[7]

Materials
20–30 ml frozen *nuclear extract* from approximately 1–5×10^9 cells or 10–20 l HeLa cells (Section 3.1.1).
$MgCl_2$ (200 mM)
Isolation buffer[a]
  65 mM KCl
  15 mM NaCl
  10 mM HEPES-KOH, pH 7.6
  10 mM EDTA
  5 mM DTT
  5 mM KF

5 mM β-glycerophosphate
0.2 µM PMSF
2 mg/ml aprotinin
Buffer D[a]
  20 mM HEPES-KOH, pH 7.9
  100 mM KCl
  0.2 mM EDTA
  5% glycerol
  0.5 mM DTT
  0.5 mM PMSF
Dialysis buffer[a]
  65 mM KCl
  15 mM NaCl
  10 mM HEPES-KOH, pH 7.6
  1 mM EDTA
  2 mM DTT
  5 mM KF
  5 mM β-glycerophosphate
  0.2 mM PMSF
$(NH_4)_2SO_4$ (fine powder)

Equipment
50 ml polypropylene centrifuge tubes (Falcon, Nunc)
SW28 rotor and Ultra-Clear^{TM} *centrifuge tubes* (Beckman)

Procedure
N.B.: All steps should be performed at 4°C.
1. Thaw *nuclear extract* (20–30 ml) (see Section 3.1.1) and dialyse for 3 h against *isolation buffer*.
2. Add *isolation buffer* to a total of 175 ml.
3. Add ammonium sulphate ($(NH_4)_2SO_4$) to 65% of saturation (0.43 g/ml supernatant).[b]
4. Stir for 2 h at 4°C.
5. Transfer to four chilled *50 ml polypropylene centrifuge tubes* and centrifuge at 8000 *g* for 20 min at 4°C.

6. Transfer supernatant to fresh tubes and centrifuge again at 8000 g for 20 min at 4°C

7. Collect supernatant and add $(NH_4)_2SO_4$ to 90% of saturation (0.183 g/ml supernatant).[b]

8. Stir supernatant for 2 h at 4°C, transfer to *Ultra-Clear*™ *centrifuge tubes* and centrifuge at 85,000 g for 1 h at (25,000 rpm in a *SW28 rotor*).

9. Gently pour supernatants out of the tubes and rinse pellet gently with 1 ml/tube ice-cold 90% $((NH_4)_2SO_4)$ saturation in *isolation buffer*—do not resuspend.

10. Resuspend pellets in a total of 5 ml of *dialysis buffer*.

11. Dialyse against three changes of 1 l of dialysis buffer over the course of 16 h.

12. Transfer dialysate to 4 fresh Eppendorf tubes and centrifuge at 13,000 g for 5 min at 4°C

13. Transfer supernatants to fresh tubes and add $MgCl_2$ to 20 mM.

14. Incubate tube for 1 h on ice and centrifuge at 13,000 g for 30 min at 4°C.

15. Gently remove supernatants and rinse pellets with 200 µl of 20 mM $MgCl_2$ in *dialysis buffer*.

16. Resuspend pellets in a total of 100 µl of *buffer D*.

17. Transfer the resuspended pellet containing the purified SR proteins to a fresh tube.

18. Check purity of SR-proteins by SDS-PAGE gel electrophoresis either followed by Coomassie staining or by immunoblotting using mAb 104 antibody.[c]

Notes

a. Add DTT and PMSF immediately before use

b. It is important to add the ammonium sulphate crystals slowly. Make sure that it is a fine powder and add only a few grams per minute.

c. Yield may vary. Generally 10–20 µg of 80–90% pure SR proteins is obtained from 1 ml of an 8–10 mg/ml concentrated nuclear extract. mAb will only detect phosphorylated SR proteins.

Comments

SR-proteins can be expressed as recombinant proteins in *Escherichia coli*, but they lack the post-translational phosphorylation of the serine residues and are poorly soluble in the absence of chaotropic reagents. It is possible to phosphorylate bacterially produced protein by preincubation in nuclear extract or, even more efficiently, by the addition of purified recombinant SR-protein specific kinases Clk/Sty[8] or SRPK.[9] Soluble and phosporylated SR-proteins can also be produced in insect cells using the baculovirus system[10] although it is not resolved whether these proteins behave exactly as those produced in human cells.

Chromatographic purification of individual SR-proteins is difficult due to their structural similarity. A SDS gel purification procedure involving a renaturation step has been published.[11]

3.2. Expression and purification of recombinant proteins in E. coli.

E. coli is the most common prokaryotic host for production of recombinant proteins provided post-translational modifications of the proteins are not required. Levels of protein expression may vary greatly from less than 1% to more than 10% of total protein in the cells. Most conventional bacterial translation vectors contain a strong *E. coli* or phage promoter, which is inducible. The initiation codon is preceded by an efficient ribosomal binding site (RBS), and convenient cloning sites are included for inserting the sequence containing the open reading frame (see Weickert et al.[12] for a review).

To ease the subsequent purification, the protein may be produced as a fusion protein, by inserting the open reading frame in frame with an affinity tag (see LaVallie and McCoy[13] for a review). Among the most commonly used affinity tags are the following: the *Schistosoma japonicum* glutathione S-transferase (GST) protein which binds glutathione, a stretch of six histidine residues binding to Ni^{2+}, *E. coli* maltose binding protein (MBP) binding to maltose, and myc- and FLAG-antigen which bind specifically to Myc- and FLAG-monoclonal antibody, respectively. The choice of expression vector may depend on properties of the protein and the nature of the subsequent experiment (see Table 3.1, for examples of commonly used protein expression vectors). Positioning of the tag at the N-terminus generally ensures efficient translational initiation, whereas the expression level of C-terminally tagged proteins is more variable. In particular the GST-tag often increases the stability and solubility of proteins and may in some cases act as a linked "chaperone" in the folding process of the fused protein. However, of particular concern is that the tag may interfere with folding and/or function of the protein. It is therefore important to check the function of the modified protein. For many experiments it is desirable to be able to radio-label the protein at specific sites, and this is readily accomplished by inserting a specific kinase recognition site in frame with the expressed protein, allowing site specific incorporation of ^{32}P or ^{33}P radioactive isotopes (see Section 3.2.4).

When producing proteins aimed for protein footprinting experiments (see Section 4.5) the purity, rather than the quantity of the expressed protein, is critical. By positioning the affinity-tag and the phosphorylation site at opposite ends of the introduced cDNA, only full-length protein becomes radiolabelled after purification[14] (see Table 4.4 in Chapter 4).

Proteins expressed in bacteria lack post-translational modifications such as phosphorylations and glycosylations. If these are crucial for function, expression in eukaryotic cells is a better choice. Procedures for expression of recombinant proteins in baculovirus or vaccinia virus can be found elsewhere.[15,16]

3.2.1. Expression of recombinant proteins in E. coli

This procedure is used routinely for expression of protein from pRSET, pGEX and pET vectors (see Table 3.1).

TABLE 3.1
Commonly used bacterial expression vectors

Plasmid/company	Tag	Matrix	Tag removal	In vitro labelling	Promoter
pRSET (Invitrogen)	(His)$_6$	Ni^{2+}	Enterokinase	No	T7
pGEX (Pharmacia)	GST	Glutathione	Thrombin factor Xa	Yes	tac
pET (Novagen)	(His)$_6$	Ni^{2+}	–	No	T7/lacO
pBAD (Invitrogen)	(His)$_6$/Myc	Ni^{2+}/Myc mAb	Enterokinase	No	Arabinose
pCAL (Stratagen)	Calmodulin binding protein (CBP)	Calmodulin	Thrombin/ Enterokinase	Yes	T7/lacO
pTYB1 (N.E.B.)	Chitin binding domain (CBD)	Chitin	Intein cleavage	No	T7/lacO
pQE (Qiagene)	(His)$_6$	Ni^{2+}	–	No	T5/lacO

Materials

E. coli cells[a]

2 × YT, 16 g/l Tryptone, 10 g/l yeast extract, 5 g/l NaCl (add 2% glucose when using GEX plasmids).

Ampicillin, 50 mg/ml

IPTG, 100 mM

Equipment

37°C incubator and a 2.5 l flask

Tubes and *rotor* for harvesting of cells

Procedure

1. Pick a single colony of *E. coli* cells and inoculate 10–50 ml of *2 × YT* medium containing 50 μg/ml *ampicillin* and let it grow overnight at 37°C with vigorous shaking.[a]
2. Dilute culture 1 : 50 into 100–500 ml of pre-warmed 2 × *YT* containing 50–100 μg/ml *ampicillin* and grow at 37°C until A_{600} = 0.8–1.0.
3. Induce protein expression by adding *IPTG* to a concentration of 0.1–1 mM and continue incubation for 2–5 h.[b]
4. Harvest cells by centrifugation at 8000 rpm for 10 min at 4°C, discard supernatant and drain pellet. The bacterial pellet can be stored at –70°C if desired. Remember to mark the original culture volume on the tube.

Notes

a. When using the T7 phage promoter, the construct must be expressed in the *E. coli* strain, which contains a genomic copy of an IPTG inducible T7 RNA polymerase, e.g. BL21(DE3), or the cells should be infected with a phage expressing the polymerase. Better yields are sometimes obtained using the modified strain BL21(DE3)-Lys which also expresses lysozyme suppressing the uninduced level of protein production.

b. The IPTG concentration and induction period should be optimised for each individual protein.

Comments
Poor protein expression and/or stability may also be overcome by:
(i) Reducing the growth temperature to 23–30°C (some bacterial
heat shock proteins are proteases); (ii) Inducing for a shorter period
of time and/or with a lower IPTG concentration. We find that
induction with 1 mM of IPTG for 30 min increases the yield and
purity of some proteins susceptible to degradation; (iii) Choosing
an appropriate *E. coli* host strain. *BL21* is a good choice due to its
deficiency in the *ompT* and *lon* bacterial proteases, but we have
found that both *XL1-Blue* and *AD202* work fine too. Addition of
nonpermeable compounds, e.g. sorbitol to the media may improve
the yield of correctly folded recombinant protein;[17] (iv) Using genes
containing codons optimised for expression in *E. coli*;[18] (v) Co-
expression of chaperones may increase the yield of correctly folded
protein.[12] See the Qiagen manual, 'the QIAexpressionist', for a
good discussion and troubleshooting guide regarding bacterial expression of proteins.

3.2.2. Purification of GST-tagged protein under native conditions

In most cases soluble GST-tagged proteins can be purified from
bacterial lysate under native conditions. The following procedure
is modified from the Pharmacia publication ('GST Gene fusion
system', 1997).

Materials
Bacterial pellet (from Section 3.2.1)
Glutathione Sepharose 50% slurry (Pharmacia)
Sonication buffer[a]
   20 mM HEPES-KOH, pH 7.9
   200 mM NaCl
   20% glycerol

1 mM EDTA
10 mM β-mercaptoethanol
0.5 µg/µl leupeptin
2 µg/µl aprotinin
200 µM PMSF
Washing buffer[a]
Sonication buffer without protease inhibitors
Storage buffer[a]
   50 mM NaCl
   50 mM Tris-HCl, pH 7.5
   1 mM EDTA
   10–20 mM reduced glutathione
Glutathione 100 mM reduced
   Store aliquated at –20°C, use within same day after thawing
Glycerol

Equipment
Sonicator
Dialysis clips and *dialysis bags* (6000–8000 MW cutoff, autoclaved
and stored in 1 mM EDTA at 4°C)

Procedure
1. Resuspend *bacterial pellet* in 50 µl ice-cold *sonication buffer* per
   ml original culture volume.
2. Sonicate cells on ice (20 s bursts followed by 10 s rest for a total
   of 4 min) and remove cell debris by centrifugation at 12,000 g
   (r_{max}) for 10 min at 4°C.[b,c]
3. Add 1 µl of a 50% *Glutathione Sepharose* equilibrated in *sonication buffer* per ml original culture volume to the supernatant
   and incubate for 30 min at 4°C with gentle agitation.
4. Collect beads by centrifugation at 1600 rpm for 3 min and wash
   5 times in 1 ml *washing buffer* (if the GST-fusion protein is going
   to be radiolabelled add a final wash in HMK buffer, see Section
   3.2.4).[d]

5. Elute the fusion protein from beads. GST-tagged proteins can be eluted shaking 3–4 consecutive times in *protein storage buffer* containing 10–20 mM reduced *glutathione* for 30 min yielding the entire fusion protein. Thrombin cleavage by 80 units of thrombin per ml of Glutathione Sepharose bed volume for 2–18 h at RT liberates the protein and leaves the GST portion bound to the Glutathione Sepharose.
6. If necessary, dialyse eluates against *storage buffer* and analyse by SDS-PAGE. Store proteins at –20°C in *storage buffer* containing 40% *glycerol*.

Notes

a. Optimal sonication, washing and storage buffer conditions will depend on the particular protein. To avoid nonspecific molecular interactions, the inclusion of a nonionic detergent or elevation of the salt concentration in the washing buffer may be beneficial.
b. The release of nucleic acids monitored spectroscopically at OD_{260} can be used as an indicator of the degree of cell lysis. The OD_{260} ought to become constant between consecutive sonications.
c. If considerable amounts of protein are insoluble and lost in the centrifugation step after sonication of the cells, the yield might be improved by the following procedure.[8] Resuspend the pellet from 1 l of cells in 8 ml of 1.5% (v/v) sodium N-lauroylsarcosinate, 25 mM triethanolamine, 1 mM EDTA (pH 8.0) and leave on ice for 30 min. Centrifuge at 18,000 g (r_{max}) for 20 min at 4°C and recover supernatant. Add 2% triton X–100 and 1 mM DTT and continue at step 3.
d. Washed fusion proteins bound to the beads can be stored at –80°C in washing buffer containing 20% glycerol.

3.2.3. Purification of His-tagged protein under denaturing conditions

The affinity of the $(His)_6$-tag for a Ni^{2+}-resin is so high that it allows contaminating proteins to be washed away under very stringent conditions. This offers the possibility of using denaturing conditions, thus allowing the purification of proteins insoluble under native conditions.[a] The following procedure is modified from the Qiagen manual ('the QIAexpressionist').

Materials
Bacterial pellet (from Section 3.2.1)
Buffer A
   6 M Guanidinium-HCl
   100 mM NaH_2PO_4
   10 mM Tris-HCl, pH 8.0 (final)
Buffer A, adjusted to pH 6.5, 5.0, 4.0 and 2.0 with HCl
Ni^{2+}-NTA resin 50% slurry (Qiagen)
Storage buffer[b]
Glycerol

Equipment
Plastic column (e.g. Biorad)
Dialysis clips and *dialysis membrane* (6000–8000 MW cutoff, autoclaved and stored in 1 mM EDTA at 4°C)

Procedure
1. Resuspend the *bacterial pellet* from Section 3.2.1 in 50 µl *buffer A* per ml original culture volume. Stir cells for 1 h at room temperature.
2. Centrifuge lysate at 10,000 rpm for 15 min at 4°C. Collect supernatant.
3. Add 2 µl of a 50% slurry of *Ni^{2+}-NTA resin*, previously equilibrated in *buffer A*, per ml original culture volume. Stir at room

temperature for 45 min, then load resin carefully into an appropriate sized *plastic column*. Adjust flow rate to 10–15 ml h^{-1}.

4. Wash with 10–15 column vol of *buffer A*, and 10 column vol of *buffer A* adjusted to pH 6.5.[c]

5. Elute the recombinant protein with 1 column vol of *buffer A* adjusted to pH 5.0, followed by 1 column vol of *buffer A* pH 4.0. Collect the protein in 0.5 ml fractions.

6. Elute with 2 column vol of *buffer A* pH 2.0.

7. Transfer fractions to *dialysis bags* and dialyse overnight against 1000 times vol of an appropriate *storage buffer*. Change buffer after 6–10 h. Store at –20°C in *storage buffer* containing 20% *glycerol*.[d]

8. Analyse fractions by SDS-PAGE.[e]

Notes

a. His-tagged proteins can also be purified under native conditions (in a procedure similar to GST-tagged proteins, Section 3.2.2), but a higher background of bacterial proteins is obtained. Elution of protein can be carried out in elution buffer containing either 100 mM EDTA or 50–500 mM Imidazole (also feasible when His-tagged protein is ^{32}P-labelled while bound to Ni^{2+}-resin).

b. Appropriate storage buffer conditions will depend on the particular protein.

c. More washing steps can be included to further remove contaminating *E. coli* proteins.

d. Expressed proteins that precipitate during dialysis may be renatured by stepwise dilution of the denaturant while still attached to the Ni^{2+}-column.

e. Fractions can be analysed by SDS-PAGE before dialysis if they are extensively diluted (>20-fold). Load the samples immediately after heating.

3.2.4. Labelling of proteins using heart muscle kinase

For many purposes it is desirable to label a protein with a radio-active isotope. In some instances, such as for protein footprinting (Section 4.5), it is crucial that the labelling takes place at a specific site in the protein. Fusing the recognition site (RRASV) for the catalytic subunit of cAMP-dependent heart muscle kinase (HMK) to a particular protein, enables specific labelling of the serine residue in the recognition site in the presence of γ-[^{32}P]ATP and the HMK enzyme.[20] The HMK enzyme is highly specific, and significant labelling of serine residues other than the one in the recognition sequence is rare. Fusion proteins expressed from some of the vectors shown in Table 3.1, and all of the vectors in Table 4.4, contain the HMK-specific site. To avoid the presence of free γ-[^{32}P]ATP and HMK enzyme in subsequent reactions, the ^{32}P-labelled affinity-tagged proteins are washed while attached to the affinity column. This procedure is essentially as published by Jensen et al.[21]

Material
HMK buffer
   20 mM Tris-HCl, pH 7.5
   100 mM NaCl
   12 mM MgCl
   0.1% Triton X-100
Washing buffer
   20 mM HEPES-KOH pH 7.9
   200 mM NaCl
   20% glycerol
   0.1% Triton X-100
γ-[^{32}P] ATP (3000–8000 Ci/mmol)
HMK enzyme (Sigma)[a]
EDTA, 0.5 M, pH 8.0
Imidazole, 1 M

Procedure

Labelling of GST-tagged proteins on beads

Because purification of GST-tagged proteins is carried out under native conditions it is possible to radiolabel the protein before elution.

1. Wash protein bound to 5–10 μl beads with 20 μl *HMK buffer* per μl bed volume Glutathione Sepharose and resuspend in 10 μl *HMK buffer*, containing 10 μCi γ-[^{32}P]ATP and 5 units *HMK enzyme* per μl bed volume (at 4°C).
2. Mix and incubate at RT for 30 min.[b]
3. Wash 5 times in 200–500 μl *washing buffer* to remove the HMK enzyme and unincorporated nucleotides (at 4°C).
4. Elution of radiolabelled protein is done similarly to elution of cold protein (see Section 3.2.2).

Labelling of His-tagged proteins

If the protein is purified under denaturing conditions it will be necessary to re-apply the protein to the Ni^{2+}-resin after renaturation and radio-labelling (otherwise renature on the column if possible). However, a high specific radioactivity can only be obtained when the protein is labelled in solution.

1. Incubate 1–5 μg His-tagged protein in 50 μl *HMK buffer* containing 10 μCi γ-[^{32}P] ATP and 5 units *HMK enzyme*, at RT for 30 min.[b]
2. Add 20 μl of a 50% slurry of *Ni-NTA resin*, previously equilibrated in *HMK buffer* and shake slowly at room temperature for 45 min.
3. Wash proteins bound to resin 5 times in 200 μl *washing buffer*.
4. Elute radiolabelled protein in 50 μl washing buffer containing either 100 mM *EDTA* or 50–200 mM *imidazole*.

Notes

a. The stability of the HMK enzyme during storage is batch-dependent but usually most active within the first week of resuspension. Store at 4°C.

b. If needed, the labelling reaction can also be performed at 4°C, but with lower efficiency.

3.3. In vitro *translation*

Traditionally, *in vitro* translation has been carried out in *E. coli* S30 extracts, rabbit reticulocyte lysates, wheat germ lysates, and more recently, in yeast lysates, and several manufacturers produce kits based on these systems. They often require that total RNA or *in vitro* synthesised RNA is generated in a separate step, so these systems are therefore of importance in studies of *cis-* and *trans*-acting elements in translational regulation. For many purposes, however, the coupled transcription-translation systems have now superceded the original *in vitro* systems, since they exhibit higher fidelity and provide a larger yield although they employ uncapped RNAs. Thus the coupled systems can be used for preparative purposes merely by the addition of plasmid DNA from a mini-prep, provided the reading frame under scrutiny is inserted down-stream from a bacteriophage promoter. In recent years there have been several reports employing these systems in studies of macro-molecular interactions. Moreover, the *in vitro* systems remain cru-cial in studies of protein sorting of pre-forms, since these precursors are difficult to isolate from cells or tissues.

The presented transcription-translation method is based on T7 RNA polymerase and the micrococcal nuclease-treated reticulocyte lysate as described by Jackson and colleagues,[22] and is suitable for a semi-preparative synthesis of recombinant protein. If few reac-tions on an analytical 10 µl scale are to be carried out the TnT® kit from Promega can be recommended.

Materials
Plasmid DNA (1 µg/µl)
Rabbit reticulocyte lysate[23]

[^{35}S]Methionine (15 mCi/ml; >1000 Ci/mmol)—e.g. Amersham
   Redivue 1594
Amino-acid mix (-methionine) (1 mM)
MgCl$_2$ (100 mM)
KCl (1 M)
rNTPs (14 mM, neutralised)
T 7 RNA polymerase (100 units/µl)
RNase T1 (1 unit/µl, Sankyo)
2 × SDS-PAGE loading buffer
   100 mM Tris-HCl, pH 6.8
   200 mM dithiothreitol
   6% SDS
   0.002% bromophenol blue
   20% glycerol

Procedure
1. Mix on ice:
    4 µl 1 M *KCl*[a]
    2 µl 100 mM *MgCl$_2$*[b]
    5 µl 14 mM *rNTPs*
    10 µl *amino-acid mix (-methionine)*
    5 µl [^{35}S]*methionine*[c]
    2 µl *plasmid DNA*[d]
    71 µl *reticulocyte lysate*[e]
2. Spin briefly to collect droplets and start the reaction with 1 µl
   T 7 RNA polymerase.
3. Incubate at 30°C for 60 min.
4. Add 2 µl *RNase T1* and incubate for an additional 30 min.[f]
5. Mix a 10 µl aliquot with 10 µl *2 × SDS-PAGE loading buffer*[g]
   and heat at 95°C for 5 min.[h]
6. Analyse the aliquot by SDS-polyacrylamide (29 : 1) gel electrophoresis.[i]
7. If necessary, purify the recombinant protein from the transcription-translation mixture.

Notes

a. Potassium chloride works better than potassium acetate in terms of fidelity, and the added concentration of 40 mM is considerably smaller than the optimal monovalent cation concentration of about 125 mM for translation of capped mRNAs, even when a correction is made for the contribution from the lysate (about 30 mM).

b. With a total rNTP concentration of 2.8 mM, the optimal $MgCl_2$ concentration is about 3.2 mM (approximately 1.2 mM is contributed by the lysate), so the concentration of 'free' Mg^{2+} is small. However, the optimal Mg^{2+} concentration ought to be determined by titration in each case.

c. For analytical purposes the inclusion of $[^{35}S]$methionine is mandatory, but for preparative purposes the radioactive label can be substituted with 100 μM unlabelled methionine.

d. It is unnecesary to linearise the plasmid, and mini-preps produced by an RNase-free kit such as JETSTAR supplied by Genomed work well.

e. The final concentrations of the energy-regenerating components are 10 mM creatine phosphate and 50 μg/ml creatine phosphokinase. Store the lysate in aliquots at $-80°C$ to avoid multiple freeze/thaw cycles.

f. RNase T1 degrades labelled peptidyl-tRNA that otherwise would interfere with the analysis of proteins having an apparent molecular mass of about 30 kDa.

g. The loading buffer contains a higher than usual amount of SDS to eliminate streaking during electrophoresis.

h. The 40 kDa 'ghost' band can be eliminated by avoiding the heat treatment.

i. Choose a percentage that is appropriate for the protein under study. A 10% polyacrylamide gel is a good compromise for the 20–100 kDa range.

Comments

In developing the coupled transcription-translation system the optimal conditions for *in vitro* transcription have been sacrificed for the benefit of efficient translation. However, the produced transcript, although uncapped, is able to saturate the translational apparatus. In fact, the low transcriptional efficiency may for various reasons facilitate the coupling between the two processes, thus promoting efficient protein synthesis overall.

If the synthesised protein is an enzyme, the crude transcription-translation mixture can be assayed straightaway; but in those cases where a binding protein without easily assayable activity is produced, purification is facilitated considerably if the binding protein is terminally tagged when the cDNA is inserted into a T7 promoter vector. Alternatively the synthesised protein can be internally tagged by including biotinylated-lysyl-tRNA in the transcription-translation reaction.

References

1. Dignam, J.D., Martin, P.L., Shastry, B.S. and Roeder, R.G. (1983). Eukaryotic gene transcription with purified components. Meth. Enzymol. *101*, 582–598.
2. Jamison, S.F. and Garcia, B.M. (1992). An ATP-independent U2 small nuclear ribonucleoprotein particle/precursor mRNA complex requires both splice sites and the polypyrimidine tract. Proc. Natl. Acad. Sci. USA *89*, 5482–5486.
3. Lee, K.A., Bindereif, A. and Green, M.R. (1988). A small-scale procedure for preparation of nuclear extracts that support efficient transcription and pre-mRNA splicing. Gene. Anal. Tech. *5*, 22–31.
4. Fu, X.D. (1995). The superfamily of arginine/serine-rich splicing factors. RNA *1*, 663–680.
5. Valcárcel, J. and Green, M.R. (1996). The SR protein family: pleiotropic functions in pre-mRNA splicing. Trends Biochem. Sci. *21*, 296–301.
6. Zahler, A.M., Lane, W.S., Stolk, J.A. and Roth, M.B. (1992). SR proteins: a conserved family of pre-mRNA splicing factors. Genes. Dev. *6*, 837–847.
7. Crispino, J.D., Blencowe, B.J. and Sharp, P.A. (1994). Complementation by SR proteins of pre-mRNA splicing reactions depleted of U1 snRNP. Science *265*, 1866–1869.

8. Gui, J.F., Tronchere, H., Chandler, S.D. and Fu, X.D. (1994). Purification and characterisation of a kinase specific for the serine- and arginine-rich pre-mRNA splicing factors. Proc. Natl. Acad. Sci. USA *91*, 10824–10828.

9. Colwill, K., Pawson, T., Andrews, B., Prasad, J., Manley, J.L., Bell, J.C. and Duncan, P.I. (1996). The Clk/Sty protein kinase phosphorylates SR splicing factors and regulates their intranuclear distribution. EMBO. J. *15*, 265–275.

10. Wu, J.Y. and Maniatis, T. (1993). Specific interactions between proteins implicated in splice site selection and regulated alternative splicing. Cell *75*, 1061–1070.

11. Zahler, A.M., Neugebauer, K.M., Lane, W.S. and Roth, M.B. (1993). Distinct functions of SR proteins in alternative pre-mRNA splicing. Science *260*, 219–222.

12. Weickert, M.J., Doherty, D.H., Best, E.A. and Olins, P.O. (1996). Optimization of heterologous protein production in *Escherichia coli*. Curr. Opin. Biotechnol. *7*, 494–499.

13. LaVallie, E.R. and McCoy, J.M. (1995). Gene fusion expression systems in *Escherichia coli*. Curr. Opin. Biotechnol. *6*, 501–506.

14. Jensen, T.H., Jensen, A. and Kjems, J. (1995). Tools for expression and purification of full length, N- or C-terminal ^{32}P labeled protein, applied on HIV-1 Gag and Rev. Gene *162*, 235–237.

15. Leffers, H. (1994). Expression of recombinant proteins using vaccinia virus vectors. In: Cell Biology: A Laboratory Handbook, Vol. 3 (Celis, J.E., ed.), Academic Press, San Diego, pp. 155–163.

16. King, L.A., Mann, S.G., Lawrie, A.M. and Possee, R.D. (1994). Baculovirus expression vector system: Production and isolation of recombinant viruses. In: Cell Biology: A Laboratory Handbook, Vol. 3 (Celis, J.E., ed.), Academic Press, San Diego, pp. 148–154.

17. Georgiou, G. and Valax, P. (1996). Expression of correctly folded proteins in *Escherichia coli*. Curr. Opin Biotechnol. *7*, 190–197.

18. Gouy, M. and Gautier, C. (1982). Codon usage in bacteria: correlation with gene expressivity. Nucleic Acids Res. *10*, 7055–7074.

19. Suñé, C. and García-Blanco, M. (1995). Transcriptional trans activation by human immunodeficiency virus type-1 Tat requires specific coactivators that are not basal factors. J. Virol. *69*, 3098–3007.

20. Edelman, A.M. (1987). Protein serine/threonine kinases. Ann. Rev. Biochem. *56*, 567–613.

21. Jensen, T.H., Leffers, H. and Kjems, J. (1995). Intermolecular binding sites of HIV-1 Rev protein determined by protein footprinting. J. Biol. Chem. *270*, 13777–13784.

22. Craig, D., Howell, M.T., Gibbs, C.L., Hunt, T. and Jackson, R.J. (1992).

Plasmid cDNA-directed protein synthesis in a coupled eukaryotic *in vitro* transcription-translation system. Nucleic Acids Res. *20*, 4987–4995.

23. Jackson, R.J. and Hunt, T. (1983). Preparation and use of nuclease-treated rabbit reticulocyte lysates for the translation of eukaryotic messenger RNA. Meth. Enzymol. *96*, 50–74.

CHAPTER 4

Preparation and analysis of RNA-protein complexes *in vitro*

4.1. Formation of RNA-protein complexes

When RNA-protein complexes are reconstituted from purified components the optimal procedure depends mainly on the participating proteins. Proteins purified under native conditions do not normally impose problems. However, many RNA binding proteins purified from cell extracts or expressed by recombinant means are isolated under denaturing conditions (urea, guanidinium or low pH), and various elaborate procedures have been developed for proper folding of denatured proteins. A simple strategy to avoid or reduce aggregation of the proteins in the binding buffer is addition of nonionic detergents.

The binding buffer should contain Mg^{2+} and sufficient monovalent ions to ensure specific binding. Divalent ions, in particular Mg^{2+}, play a critical role in proper tertiary folding of RNA. In general, complex RNA such as rRNA requires 10 mM $[Mg^{2+}]$ whereas less complex molecules such as most mRNAs require a lower Mg^{2+} concentration to fold properly. The monovalent ion concentration should be about 100–300 mM. KCl or KOAc is generally used as the main source of monovalent ions to simulate the intracellular conditions.

Procedure

The optimal conditions should be determined for each RNA-protein complex.

1. The first step is renaturation of the RNA (Section 2.1.2.6).

2. Cool the renatured RNA to approximately 30°C (for some complexes 0°C).
3. Adjust the K$^+$ concentration.
4. Add (if necessary) Nonidet (P-40) or Tween 20 to 0.01%.
5. Add the protein.[a]
6. Incubate for 30 min at 30–37°C.[b]

Notes

a. Ideally, the protein should also be renatured by a pretreatment at 30°C before addition to the RNA. However, many proteins purified under denaturing conditions tend to aggregate during the renaturation procedure. Occasionally, aggregation in the protein renaturation buffer can be reduced by addition of carrier RNA.
b. The temperature depends on the specific protein.

4.2. Analysis of RNA-protein complexes

Evaluation of protein-RNA complex formation is important in many studies. In this section, three commonly used methods are described; filter binding, mobility shift and sucrose gradient centrifugation. All three methods are nonequilibrium methods where the complex is formed (at equilibrium) at high concentrations of the components. However, during the subsequent separation the equilibrium situation is distorted, since the association of the RNA and the protein is hampered by their physical separation, so only dissociation of the complex can occur. Therefore, the choice of technique depends critically on the association and dissociation kinetics of the complex. In experiments with purified components filter binding assays will often be fast enough to study rather labile complexes, while mobility shift assays and sucrose gradient centrifugation can be employed for more stable complexes.

Binding of proteins to RNA often induces conformational

changes in the RNA that alters the spectroscopic properties of the RNA such as ellipticity. Circular dichroism can then be used at equilibrium to measure the binding. Recent developments in plasmon surface resonance absorption methods (Biacore, Pharmacia) have made this technology, which is independent of changes in spectroscopic parameters, an attractive approach for more accurate determinations of binding properties. However, we will not discuss these methods in detail since it requires rather expensive and specialized equipment.

4.2.1. Filter binding assay

The filter binding technique is a rapid and a commonly used technique for detecting protein-RNA binding and evaluation of binding constants. The principle behind filter binding assays is the ability of nitrocellulose filters to retain proteins, and ideally also the bound RNA, while the naked RNAs pass through the filter. Despite its conceptual simplicity, reliable analysis of RNA-protein complexes is technically demanding and unsuitable for some complexes. The assay works well for relatively small RNA molecules, while large RNA molecules in complex are often not retained.

The assay is performed by mixing trace amounts of radioactively labelled RNA with protein at different concentrations and then filtering through a nitrocellulose filter. Ideally, 100% of the RNA should be retained at high protein concentrations, but in reality a lower retention efficiency (percent RNA retained in the presence of a large excess of protein) is obtained. In addition, a fraction of the protein might not be retained by the filter. As an example, filter binding analysis of the 5 S rRNA-L18 complex revealed that the retention efficiency was 35% for the 5 S rRNA-L18 complex while 65% of the L18 protein was retained and <5% of free 5 S rRNA was retained.[1] However, the accuracy of the measurements does not depend directly on the retention efficiency, provided a constant proportion of the complexes are retained at the concentrations tested.

A number of parameters influence the retention efficiency of the assay and should be optimized for each binding protein studied:

Buffer: Increasing salt concentrations lower the retention efficiency. For most complexes the monovalent ion concentrations should be above 150 mM to ensure specificity of binding.

Temperature: Low temperature increases the retention efficiency.

Flow rate and washing conditions: The flow rate should be kept constant, i.e. apply the same suction in each experiment. Unspecific binding of naked RNA to the filter can be reduced by extensive washing, but often at the expense of the retention efficiency. It is for some complexes an advantage with consecutive (3–4) washes with small volumes, while for other complexes the best results are obtained when the complex is applied in a larger volume.

RNA renaturation: It is well known that different RNA preparations give variable retention efficiencies. Careful renaturation often reduces the variability between different preparations (see Section 2.1.2.6).

Materials
Nitrocellulose filters (Millipore, type HA)
Preparation of the filters
  Soak and degas the nitrocellulose filter in the binding buffer
Binding buffer[a]
  30 mM Tris-HCl, pH 7.8
  10 mM MgCl$_2$
  150 mM KCl

Equipment
Filtration unit (connected to constant pressure vacuum unit).

Procedure
Prepare the protein-RNA complex as described in Section 4.1 using radioactively labelled RNA. Clear the solution by centrifugation for 1 min at 13,000 *g*, and keep the precipitate for scintillation counting.

1. Place the degassed filter in the filtration unit.
2. Add binding buffer and apply vacuum. Use flow rate of 0.5–
   1 ml/min (be careful not to dry the filter).
3. Dilute the cleared solution with the complex in 100 μl binding
   buffer.[b]
4. Add the solution to the washed filter.
5. Wash the filter 2–5 times each with 100 μl buffer[b,c] (let the
   binding buffer flow through after each addition without drying
   the filter).
6. Dry the filter, add scintillation fluid and count in a scintillation
   counter.

Notes
a. The binding buffer should be optimized for each complex.
b. It should be examined whether dilution of the RNA-protein
   complex or repetitive washes gives the optimal retention efficiency.
c. The whole filtration procedure should be finished in less than
   1 min.

Comments
If only trace amounts of RNA are used, the K_d can be estimated
from the following equation:

$$\frac{R_b - R_0}{R_{tot}} = R_e \frac{[protein]}{K_d + [protein]}$$

where R_e is the retention efficiency, R_b counts bound in the presence of protein, R_0 counts bound in the absence of protein, and
R_{tot} total counts of RNA applied to the filter (corrected for the
precipitated counts).

To measure the stoichiometry of the RNA-protein complex a
similar assay can be used but this time with stoichiometric amounts
of RNA and protein (under these conditions [protein] in the

equation must be corrected for protein associated with the RNA, i.e. [protein] = [total protein] – [RNA-protein complex]). If the results indicate association of more than one protein per RNA molecule it might imply that unspecific binding or only a fraction of the protein is active in binding. The filter assay cannot distinguish between these possibilities, but a mobility shift analysis may be able to resolve the problem.

These limitations, and the fact that nitrocellulose-complex inter-action might significantly interfere with complex stability, should be considered before detailed thermodynamic parameters are derived from the experiments.

4.2.2. Mobility shift analysis

The mobility shift assay (or gel retardation assay) is based on the difference in mobility between naked RNA and proteins-RNA complexes when migrating in a native polyacrylamide gel. The increased size of the complex and the usually low or positive charge of nucleic acid binding proteins give rise to the difference in mobility.

The qualitative interpretation of the mobility shifts is straightforward for purified proteins, although more quantitative assessments should be performed with care (see Comments).

The method enables a direct evaluation of the stoichiometry of the protein-RNA complex. If more than one protein binds, additional shifts (supershifts) in the RNA mobility can be observed (Fig. 4.1). Supershifts after addition of antibodies can also be used to identify proteins in the complex when crude preparations of protein are used for complex formation.

A major concern in mobility shift assays is to select conditions which ensure specific binding. In general this is accomplished by high concentrations of salt (300–500 mM NaCl or KCl). Unfortunately, high salt concentrations frequently increase the rate of dissociation of the complex during the electrophoresis. Thus the salt concentrations may have to be optimized depending on the particular experiment. Many mobility shift experiments with protein-RNA

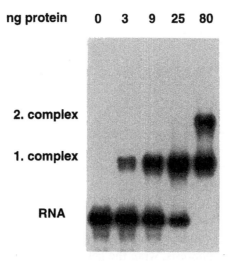

Fig. 4.1. Mobility shift analysis: Mobility shift analysis performed by 4% PAGE in 50 mM Tris-glycine 1 mM EDTA. The mobility shift is induced by the 60 kD polypyrimidine tract binding protein to a 220 nucleotide random labelled pre-mRNA transcript. The experiment shows dimeric binding (2. complex) of the protein at high concentration.

interactions have been studied in 50 mM Tris-glycine buffers while specific *Escherchia coli* ribosomal protein-rRNA complexes require at least an additional 150 mM Na^+ and 5 mM Mg^{2+} to ensure specific interaction.

If the complex is sensitive to increased salt concentrations the nonspecific binding can be reduced by addition of carrier RNA.

Materials
Native gels
  Acryl amide: 4–8% (acrylamide : bis acrylamide; 39 : 1)[a]

  Gel buffers:
    pH buffer: 80 mM Tris-glycine, pH 8.8, or Tris-Acetate, pH 7.8, or Tris-Borate, pH 8.0[b]

Divalent ions: 1–10 mM Mg(OAc)$_2$[c]
Monovalent ions: 0–300 mM NaOAc or KOAc[d]

Load buffer
  80% glycerol

Procedure
1. Prepare the RNA-protein complex as described in Section 4.1 using radioactively labelled RNA.
2. After incubation of the RNA and protein, spin the solution for 1 min at 13,000 *g*.
3. Transfer the supernatant to a new tube and add 1/10 vol *load buffer*.
4. Load the gel (after a short prerun).[e]
5. After electrophoresis, wrap the gel in plastic foil and autoradiograph.[f]

Notes
a. For large complexes a lower cross-linking ratio (e.g. 60 : 1 or 80 : 1 (Section 5.1.3) is preferable. Optional addition of 0.5% melted agarose may be beneficial for the low cross-linking ratios.
b. The selection of anion depends on the stability of the complex in the different buffers. Cl$^-$ should be omitted since chlorine will develop during the electrophoresis.
c. The optimal Mg^{2+} concentration depends on the complexity of the RNA. Simple hairpin structures generally require less Mg^{2+} to fold properly than, e.g. RNA junctions.
d. Higher concentration of monovalent ions increase the heat generation during the electrophoresis.
e. Load dyes in a separate slot since they may interfere with the complex.
f. The amount of RNA in the bands can either be measured by excising the bands and counting them in a scintillation counter or by using a PhoshorImager system.

Comments

Quantitative assessments based on mobility shift experiments are questionable. A simple evaluation of the parameters illustrates the problem. The kinetics of the complex formation can be described as;

$$\frac{d[\text{Complex}]}{dt} = k_{\text{on}}[\text{RNA}][\text{Prot}] - k_{\text{off}}[\text{Complex}]$$

Since the RNA is only in trace amounts compared to the protein, the assembly of the complex will initially follow pseudo first-order kinetics,

$$k_{\text{pso}} = k_{\text{on}}[\text{Prot}] = k_{\text{off}}\frac{[\text{Prot}]}{K_{\text{d}}}, \quad \text{since} \quad K_{\text{d}} = \frac{k_{\text{off}}}{k_{\text{on}}}$$

For most complexes k_{pso} is in the order minutes. Thus, under conditions where the protein concentration [Prot] is equal to K_{d} (the protein concentration which in the mobility shift assay produces equal amounts of RNA in the free and the bound form) $k_{\text{off}} = k_{\text{pso}}$, i.e. approximately 1 min! Since a mobility shift experiment takes at least 1 h, no complex should be left. There might be different explanations to this enigma; the affinity measured from the assay might be an underestimate of the real value, or the low salt conditions generally used in the gel might change the kinetic parameters radically so that the complex in the gel might be trapped in a kinetically stable form due to low salt. When the salt concentration is raised in the gel the complex becomes more labile, and the RNA dissociated from the complex appears as a smear between the bands originating from the intact complex and the free RNA.

4.2.3. Sucrose gradient analysis

Preparative rate-zonal centrifugation of ribonucleoprotein particles
in preformed sucrose gradients is less sensitive and exhibits lower
resolution than electrophoretic approaches, but the capacity is considerably higher and the choice of ionic conditions far more wide-
spread. Although it is possible to form gradients in various shapes
to meet special demands such as an isokinetic gradient for
determination of sedimentation coefficients, the most popular type
of gradient is the linear sucrose gradient. Since ribonucleoprotein
complexes are more dense than the highest possible sucrose density
of 1.3 g/ml they will always sediment to the bottom of the tube
given sufficient time, so the choice of gradient is dictated by the
separation at hand; i.e. a 5–20% (w/w) gradient is typically chosen
for separations of smaller ribonucleoproteins, a 10–30% (w/w)
gradient for separations of spliceosomes, ribosomal subunits and
monosomes, whereas a 20–47% (w/w) gradient may be chosen if
emphasis is on the separation of individual polysome sizes. A steep
5–56% (w/w) gradient is a compromise where some resolution of
all species is obtained without pelleting the large particles. Here
we present an 'unorthodox' method of preparing and fractionating
sucrose gradients in conjunction with ultracentrifugation in the popular long-bucket swing-out Beckman SW41 rotor, which we have
found useful in the preparation and analysis of various ribonucleoprotein particles. In the Notes section, following the basic protocol,
there are brief descriptions of more conventional approaches to
preparing and fractionating sucrose gradients.

Materials
Light sucrose solution
  10% (w/w) sucrose (Serva)
  20 mM Tris-HCl, pH 8.0
  5 mM $MgCl_2$
  140 mM KCl

Dense sucrose solution
   30% (w/w) sucrose (Serva)
   20 mM Tris-HCl, pH 8.0
   5 mM $MgCl_2$
   140 mM KCl

Equipment
Polyethyleneterephthalate or polyallomer tubes
Ultracentrifuge
Swing-out bucket rotor (e.g. Beckman SW41 Ti rotor—*k** factor: 355)
Gradient-mixer of the two-cylinder type
Peristaltic pump
UV spectrophotometer

Procedure
1. Close the valve between the chambers and the outlet of the *gradient mixer*, and pour 6 ml *light sucrose solution* into the mixing chamber where there is a magnetic stirring bar. Open the valve between the chambers briefly to allow a drop to enter the reservoir chamber. Pour 6 ml *dense sucrose solution* into the reservoir. Place the *gradient mixer* on a magnetic stirrer (Fig. 4.2).[a]
2. Connect the outlet of the *gradient mixer* to a *peristaltic pump* (multichannel if more than one gradient is prepared), and attach a glass capillary longer than 10 cm to each tubing outlet. Place the glass capillary in the bottom of the *ultracentrifuge tube*.[b]
3. In the following order: Start the magnetic stirrer, open the gradient mixer outlet, open the valve between the chambers, and start the peristaltic pump. Fill the tubes and take care to stop the pump before air is introduced into the bottom of the tube. Withdraw the capillary carefully.
4. Place the filled tubes in prechilled SW41 buckets and leave on ice while the sample is prepared.[c]
5. Apply the sample gently to the meniscus of the gradient.[d] Weigh

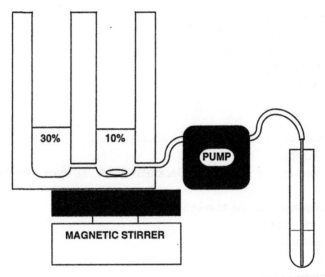

Fig. 4.2. Schematic illustration of a sucrose gradient mixing device.

opposing buckets and add additional sample buffer if necessary and screw the aluminium caps on.[e]

6. Hook all buckets onto the chilled *SW 41 rotor* and place it in the ultracentrifuge chamber of 4°C. At the base of the rotor there are marks that indicate the position of the pins at the bottom of the rotor, so they can be placed perpendicularly to the pins on the drive.

7. Accelerate slowly up to 100 g (1000 rpm in SW41 Ti rotor) and centrifuge at 197,000 g at r_{av} (40,000 rpm in SW41 Ti rotor) and 4°C for 4 h.[f] Decelerate from 1000 rpm without braking.

8. Gently insert a glass capillary at the bottom of the tube and pump out the gradient through a $UV_{260 \text{ nm}}$ *spectrophotometer* letting the gradient in at the top of the flow-cell and collecting 1 ml fractions from the bottom of the flow-cell.[g]

9. Analyse fractions.[h]

Notes

a. Usually more than one gradient is prepared at a time, in which case the volumes are multiplied by the relevant factor. The valve between the chambers are opened briefly to dispel trapped air. In the more conventional setup of filling the ultracentrifuge tubes from the top, the dense sucrose solution is placed in the mixing chamber and the light sucrose solution is placed in the reservoir chamber.

b. Filling the tubes from the bottom is preferable with nonwettable tubes of the polyallomer type.

c. The rotor should be kept in the refrigerator overnight before the experiment. Remember to use opposing buckets, i.e. 1 and 4, 2 and 5, and/or 3 and 6.

d. The sample concentration expressed in percent w/w should not exceed one tenth of the starting gradient concentration, and a volume of 500 µl is regarded as the maximum sample size (<4% of the gradient volume) for the SW41 tube. The gradient is able to handle up to about 30 µg material/ml gradient before being overloaded.

e. The aluminium caps are soft (and expensive!), so take care that the thread is not damaged.

f. It is common to use the maximum speed of the rotor (*Note*: some rotors may have to be derated due to wear or to sucrose concentrations in excess of 47% (w/w), but keep in mind that high hydrostatic pressure can dissociate complexes. The employed conditions correspond to a particle of about 50S sedimenting halfway through the gradient.

g. This is an unconventional way of fractionating a gradient but it eliminates the need for special equipment in connection with the upward displacement procedure that introduces a dense solution at the bottom of the tube. To avoid 'pulsation' in the flow-cell due to gradient inversion, the gradient is led in at the top of the flow-cell and out at the bottom. In this approach, the flow-cell should be flushed with the dense sucrose solution, before the gradient is pumped through.

h. In a new series of experiments a small aliquot of each fraction should be withdrawn for refractometry (50 μl is sufficient for an Abbé type refractometer) to establish the density profile of the gradient. The density in g/ml at 0°C can be calculated from the equation $\rho = 2.7329\eta - 2.6425$ where η is the refractive index at 20°C corrected for the presence of solutes other than sucrose.

Comments

The prevalent use of sucrose gradients in rate-zonal centrifugation over the years means that there are protocols for the use of sucrose gradients for many types of separation. Therefore, a comparison between $\omega^2 t$-values for different runs with the same gradient and rotor will usually suffice. However, it should be recalled that among the linear sucrose gradients it is only the 5–20% (w/w) gradient that is essentially isokinetic. Additional information regarding sucrose gradient analysis can be found in Owen Griffith's booklet[2] and in The Practical Approach Series.[3]

4.3. Isolation and identification

4.3.1. Affinity purification using tagged RNA or protein

A direct method for testing RNA-protein interaction *in vitro* is by affinity-precipitation or -chromatography. Two molecules interact if specific purification of one of them leads to co-purification of the other molecule. This approach is often referred to as the 'pull-down method'. Different experimental designs can be used: One possibility is to let the RNA-protein complex form in solution under native conditions, and then purify the complex by binding a high affinity tag, present in either the RNA or the protein to beads. The physical separation may be accomplished by low speed centrifug-ation (batch approach) or by column chromatography (column

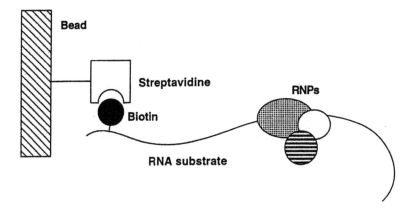

Fig. 4.3. Affinity purification of RNA associated complexes. The RNA may be biotinylated during transcription and complexes are allowed to form in solution. The complexed RNA can then be immobilised on streptavidin coated beads and purified. Associated molecules can be eluted from the RNA and analysed by the method of choice. Variations of this scheme are described in the text. In certain cases it may be favourable to form the complexes after the immobilisation step.

approach) (Fig. 4.3). Alternatively, either the RNA or the protein is immobilised prior to complex formation. The latter method suffers from being slower in the complex formation step due to the kinetical constraints on one of the components, but is advantageous if one of the components is dilute or if a large volume has to be assayed.

Since many proteins exhibit some unspecific binding to nucleic acids, it is crucial to include appropriate controls. In investigations of protein binding to immobilised RNA a control in which the RNA is replaced with an unrelated or mutated RNA is often used.

The specific binding of RNA or proteins to beads can be accomplished by various methods. RNA may be tethered to a matrix by incorporating biotin, a 244-Dalton vitamin, randomly in the RNA during transcription (see Section 2.3.2) so that the RNA subsequently can be attached to an avidin- or streptavidin column (see Section 4.3.1.1). Another approach utilises a biotinylated anti-sense oligonucleotide made of either DNA, RNA, or 2′-O-modified RNA, complementary to an accessible region of the RNA as an adapter between the RNA and the avidin- or streptavidin column. The latter approach has the advantage that the RNA-protein complex is more easily removed from the matrix and it eliminates the possibility that some RNA-protein complexes may be inhibited by the presence of biotin in the RNA. However, the target region within the RNA must be readily accessible to obtain sufficient annealing of the antisense oligonucleotide, and care most be taken that no RNase H activity is present if DNA oligonucleotides are employed. This is a particular concern when using nuclear extract which both contain a large number of general RNA binding proteins and RNaseH activity. For this purpose, the nuclease resistant 2′-O-alkyl RNA oligonucleotides are the best choice (see Ref. 4 for a detailed procedure).

Binding of native proteins to beads may also occur through the biotin-avidin interaction (see Section 4.3.1.1) but several alternative approaches are widely used. Proteins may be expressed as a fusion-protein containing a convenient tag-peptide. Commonly used tags include glutathion-S-transferase protein (GST-tag) that binds to glutathion-coupled beads (Section 4.3.1.2), histidine tag (His-tag) composed of 6–8 histidines in a row that specifically interacts with Ni^{2+}-tethered beads (Section 4.3.1.2), maltose binding protein (MBP-tag) binding maltose coupled beads, and a number of different immuno-tags which will interact with a specific antibody immobilised on the beads.

Coupling of a nontagged protein to a matrix is also possible using a number of different chemicals reactive towards amino acid residues. However, coupling of the protein to the matrix generally requires relatively harsh chemical conditions, and the method is therefore only suitable when the protein can be coupled to the matrix prior to the complex formation. A variety of procedures for chemical coupling of proteins to a matrix are given in a comprehensive handbook from Pharmacia (Affinity Chromatography, Principles and Methods, Code number 18–1022–29) and will not be covered in detail here. The procedures below are adjusted for analytic purposes, so for preparative purposes the procedures should be scaled up accordingly.

4.3.1.1. Binding of biotinylated protein or RNA to streptavidin beads

One of the most popular approaches for binding of macromolecules to a solid support utilises the high affinity of the vitamin biotin to avidin and streptavidin which is one of the strongest known noncovalent biological recognition between protein and ligand (approximate $K_a = 10^{15}$ M^{-1}). Due to the relatively small size of biotin it can often be covalently attached to RNA and protein without altering the biological activity. Biotin is usually incorporated into RNA during *in vitro* transcription (see Section 2.3.2), whereas biotinylation of proteins is carried out post-translationally (see Note a).

Materials
Biotinylated RNA (see Section 2.3.2 for preparation) or *protein*[a]
Streptavidine-agarose beads (BRL)[b]
Blocking buffer
  20 mM HEPES-KOH, pH 7.6
  300 mM KCl
  0.01% Nonidet P-40
  E. coli bulk tRNA (0.1 mg/ml; RNase free from Boehringer Mannheim)
  0.5 mg/ml BSA
Tag-binding buffer
  20 mM HEPES-KOH pH 7.6
  100 mM KCl
  0.01% Nonidet P-40
  E. coli bulk tRNA (0.1 mg/ml; RNase free from Boehringer Mannheim)
  0.5 mg/ml BSA

Washing buffer
  20 mM HEPES-KOH pH 7.6
  100 mM KCl
  0.01% Nonidet P-40
Binding buffer (see Section 4.3.1.3)

Equipment
Rotating wheel stirrer

Procedures
Preblocking streptavidine-agarose beads:
The beads prepared in this blocking procedure will suffice for 4–5 analytical affinity purification experiments.

1. Add 100 μl of a 50% suspension of *streptavidin-agarose beads* to an Eppendorf tube.[b]
2. Add 10 volumes of *washing buffer*.
3. Collect beads by centrifugation for 30 s at 1500 *g*.[c]
4. Remove the buffer—be careful not to remove any beads.
5. Repeat the washing steps (steps 2–4).
6. Rotate beads with 4 bed volumes of *blocking buffer* at 4°C for 30 min.
7. Centrifuge 30 s at 1500 *g* and remove buffer.
8. Repeat the washing steps (step 2–4) twice.
9. Add 100 μl *washing buffer* and store on ice. The beads are now preblocked and ready for use (use within 1–2 h).

Binding biotinylated molecules to beads
This procedure describes a method for preparing RNA or protein attached to beads aimed for subsequent affinity purification experiments. To bind a biotinylated RNA-protein complex present in a cellular extract, see Section 5.1.6.

1. Mix in 100 μl volume:
    1 μg of the *biotinylated RNA* or *biotinylated protein*.
    20 μl of the 50% suspension of preblocked beads prepared as described above.

Adjust volume with *tag-binding buffer*.
2. Rotate at 4°C for 1 h.
3. Collect beads by centrifugation for 30 s at 1500 *g* and resuspend in 10 volumes of appropriate *binding buffer* (see Note a in Section 4.3.1.3).
4. Collect beads by centrifugation and resuspend in 20 µl of the *binding buffer*, and go to Section 4.3.1.3.

Notes
a. Biotinylation of a protein can be carried out by mixing 1 mg of protein in 0.5 ml of 50 mM sodium bicarbonate buffer, pH 8.5 with 50 µl of a 1 mg/ml freshly dissolved solution of Sulfo-NHS-LC-Biotin (Pierce) in water, followed by a 2-h incubation at 0°C or 30 min at room temperature. Unreacted Sulfo-NHS-LC-Biotin is removed by dialysis or gel filtration. Sulfo-NHS-LC-Biotin will primarily react with the primary amines at these conditions. Since the reactivity of various proteins differs considerably titrations with different concentrations of the biotin reagent should be performed. Biotin compounds reactive towards other amino acid side-chains are available from Pierce.
b. Alternative streptavidin-coated iron beads from Dynabeads™ (Dynal) can be used which can be harvested using a magnet.
c. Centrifugation of beads at higher speeds can damage them. This also applies to all the washing steps.

4.3.1.2. Binding of protein to glutathione beads
Materials
Glutathione Sepharose (50% slurry; Pharmacia)
GST-tagged protein 1 µg
Tag-binding buffer
  20 mM HEPES-KOH pH 7.6
  100 mM KCl
  0.01% Nonidet P-40
  E. coli bulk tRNA (0.1 mg/ml; RNase free from Boehringer Mannheim)

0.5 mg/ml BSA
Binding buffer (see Note a in Section 4.3.1.3).

Equipment
Rotating wheel stirrer

Procedure
The beads are first blocked with unspecific protein and RNA as described for streptavidine-agarose beads in Section 4.3.1.1.
1. Mix in 100 µl volume:
   1 µg of *GST-tagged protein* in *tag-binding buffer*.
   20 µl of the 50% suspension of *Glutathione Sepharose resin*.
   Adjust volume with *tag-binding buffer*.
2. Rotate at 4°C for 1 h.
3. Collect beads and resuspend in 10 volumes of appropriate *binding buffer*.
4. Collect beads, resuspend in 20 µl of the *binding buffer* and go to Section 4.3.1.3.

Comments
The same procedure is used for binding of His-tagged proteins to Ni-NTA, except the binding buffer must not contain DTT and EDTA. DTT may be replaced with 10 mM β-mercaptoethanol.

4.3.1.3. Affinity purification of RNA/protein
The immobilised RNA or protein can be used as a bait to isolate a specific ligand from a complex pool of macromolecules. When using immobilised protein the Ni-NTA resin generally yields a higher background of unspecific binding than the strepavidine and the glutathione resins. We recommend to use a glutathione sepharose matrix and GST-fusion protein as the first choice and alternatively, if the protein is nonfunctional with the GST-tag, use the biotin-streptavidine system. Remember to avoid EDTA and DTT in all buffers if Ni-NTA is used to immobilise the complexes.

Materials

Beads containing immobilised protein or RNA (50% slurry in *Binding buffer*[a]) prepared as described in Section 4.3.1.1 or 4.3.1.2.

Binding buffer[a]

10 mM HEPES-KOH, pH 7.9

100 mM KCl

2 mM $MgCl_2$, 0.5 mM EDTA

1 mM DTT

10% glycerol

0.2% Tween 2

RNasin (100×; 40 units/μl; Promega)

E. coli tRNA (100×; 5 μg/μl; RNase free; Boehringer Mannheim)

Bovine serum albumin (50×; 10 μg/μl)

RNA elution buffer

0.3 M NaOAC (pH 6.0)

1 mM EDTA

50 ng/μl *E. coli* tRNA

Protein elution buffer

58 mM Tris-HCl, pH 6.8

6% glycerol

1.7% SDS

0.0025% Serva Blue W

0.8% β-mercaptoethanol

RNA loading buffer

80% formamide, v/v

0.1% xylene cyanol

0.1% bromophenol blue

1 mM EDTA, pH 8.0

Analysing RNA bound to immobilised protein

1. Mix:

20 μl 50% slurry *beads containing immobilised protein*.

50–200 μl *binding buffer* containing 0.4 units/μl *RNasin* (Promega), 50 ng/μl *E. coli* tRNA, 0.2 μg/μl *bovine serum albu-*

min and the RNA partner (alone or within a complex mixture of molecules).

2. Rotate at 4°C for 5–30 min.[b]
3. Collect beads by centrifugation for 30 sec at 1500 *g*, and wash beads five times in 800 μl *binding buffer* (without RNasin, bovine serum albumin and tRNA).[c]
4. Add 100 μl of *RNA elution buffer* and extract once with *phenol* and once with *chloroform*. Add 2.5 vol of ethanol and precipitate RNA at −70°C for 15 min.
5. Pellet the RNA and resuspend in *RNA loading buffer*, and incubate at 95°C for 3 min prior to gel analysis.

Analysing protein bound to immobilised RNA
1. Mix:
   20 μl 50% slurry *beads containing immobilised RNA*.
   50–200 μl *binding buffer* containing 0.4 units/μl *RNasin* (Promega), 50 ng/μl *E. coli* tRNA, 0.2 μg/μl *bovine serum albumin* and the protein partner (alone or within a mixture of molecules).
2. Rotate at 4°C for 5–30 mins.[b]
3. Wash beads five times in 800 μl *binding buffer* (without RNasin, bovine serum albumin and tRNA).
4. Retained proteins are released by incubating the beads in 40 μl *protein elution buffer* at 95°C for 5 min.
5. The samples (avoid the beads) are generally subjected to SDS-PAGE so proteins can be visualised by Coomassie or silver staining, or autoradiography if the protein is labelled.[d]

Notes
a. This is a general binding buffer which may work in many systems. However, the conditions may vary depending on the system and should therefore be optimised individually. Adjust the salt (generally KCl or NaCl), nonionic detergent (Tween 20 or Triton X–100) and unspecific carrier RNA/protein to minimise

nonspecific ionic and hydrophobic interactions, without affecting the specific interactions investigated.

b. Depending on the on-rate kinetics for the particular RNA-protein interaction.

c. For weaker interactions it may be necessary to reduce the KCl concentration in the washing buffer. However, this will also decrease the specificity of the interaction.

d. The proteins may be transferred to a membrane followed by a western blot analysis if desired.

4.3.2. UV cross-linking

RNA-protein complexes may be covalently associated by cross-linking the participating molecules. This allows a subsequent denaturing treatment that would otherwise dissociate the components of the complex. By irradiating with ultraviolet light of 254 nm, ^{32}P-labelled RNA is cross-linked to a potential RNA-binding protein in an extract, and the approximate size of the binding protein is estimated by SDS-PAGE analysis as shown in Fig. 4.4.

Alternatively, photoreactive nucleoside analogues can be incorporated enzymatically on a random basis or at specific positions in an RNA by using the methods described in Sections 2.3.2 and 2.3.4, respectively. In this way, less detrimental irradiation conditions (longer wavelength and shorter time) can be employed to obtain the cross-link. Table 4.1. lists some photoreactive nucleoside analogues that have been incorporated into *in vitro* transcripts by T7 RNA polymerase.

Materials

[^{32}P]-RNA

In vitro synthesised RNA is labelled by the inclusion of [α ^{32}P]UTP (3000 Ci/mmol; 10 mCi/ml) in a standard 10 μl reaction as described in Section 2.3.1 and purified by denaturing gel electrophoresis or through a spin-column (see Section 2.1.2).

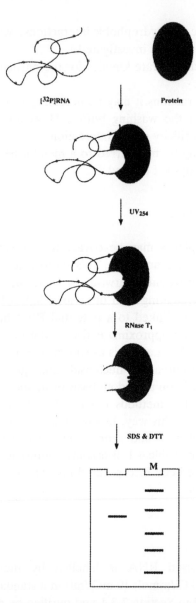

Fig. 4.4. Outline of a UV cross-linking experiment.

TABLE 4.1
Photoreactive 'zero-length' nucleoside analogues

Analogue	λ_{max} (nm)	Source	Ref.
4-thiouridine	330	4-thio-UDP (Sigma)	
5-bromouridine	280	5-bromo-UTP (Sigma)	
5-azidouridine	288	5-azido-UTP	5
5-iodouridine	290	4-iodo-UTP (Sigma)	
5-iodocytidine	295	5-iodo-CTP (Sigma)	
6-thioguanosine	342	6-thio-GTP	6

Purified RNA is redissolved at about 50 fmol/μl with 20 ng/μl *E. coli* tRNA carrier.

Crude cytoplasmic lysate

5×10^7 cells are lysed in 1 ml of lysis buffer (20 mM Tris-HCl, pH 8.5, 1.5 mM $MgCl_2$, 140 mM KCl, 1 mM DTT, 0.5% Nonidet P-40) by pipetting up and down. Nuclei, mitochondria and debris are removed by centrifugation at 13,000 *g* for 15 min, and the crude cytoplasmic lysate is made 5% in glycerol before it is stored in aliquots at −80°C.

Binding buffer

20 mM Tris-HCl, pH 8.1

5 mM $MgCl_2$

140 mM KCl

0.5 mM DTT

2% glycerol

2 × SDS-PAGE loading buffer

100 mM Tris-HCl, pH 6.8

200 mM DTT

6% SDS

0.002% bromophenol blue

20% glycerol

Heparin (2.5 units/μl—about 15 μg/μl)

RNase T1 (2.5 units/μl, Sankyo; 80 units/μl, Amersham)

RNase V1 (0.15 unit/μl, Pharmacia)

Equipment
Stratalinker 1800 or equivalent UV source

Procedure
1. Incubate 50 fmol [^{32}P]RNA[a] and 15 μg protein from a *crude cytoplasmic lysate* in 10 μl *binding buffer* at room temperature for 15 min.
2. Add 1 μl *heparin*[b] and continue incubation for a further 10 min.
3. Place the sample on ice in the bottom of a *Stratalinker 1800* and irradiate with 254 nm UV light (3 mW/cm^2) for 30 min.
4. Add 2.5 units *RNase T1*[c] and 0.15 units RNase V1[d] and incubate for 20 min at 37°C.
5. Add 12 μl *2 × SDS-PAGE load buffer*, heat at 95°C for 3 min, and apply an aliquot to the gel.

Notes
a. Renaturation of gel-purified [^{32}P]RNA in binding buffer (Section 2.1.2) prior to the addition of crude lysate is recommended.
b. The inclusion of heparin after 15 min of incubation is optional since the cross-linking pattern of many interactions does not change substantially.
c. It may be necessary to add more units of RNase T1 if it is purchased from another supplier, e.g. 80 units of the Amersham enzyme.
d. RNase V1 treatment is optional but it reduces the background when the RNA component exhibits a high degree of helical geometry.

Comments
It should be recalled that the UV cross-linking approach is a two-step procedure, so a high-affinity binding protein may pass unnoticed if the prerequisites for the photo-addition are not present. The major concern regarding the UV cross-linking approach is the

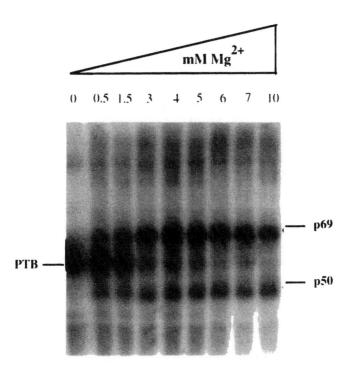

Fig. 4.5. Mg^{2+}-dependence of the identity of UV cross-linked proteins. A randomly labelled 5' untranslated region from a growth factor mRNA was mixed with a cytoplasmic lysate at increasing Mg^{2+} concentrations in the binding buffer, and UV cross-linking was carried out as described in the text. PTB is the polypyrimidine tract binding protein, p50 is the major core protein of mRNPs, and p69 is a leader-specific binding protein.

specificity of the obtained cross-link, since UV light of 254 nm pertubates RNA and protein conformations considerably and may even lead to degradation. Therefore, a high cross-linking efficiency is more likely to reflect an interaction between native conformers than a low yield which may be the result of fixating aberrant conformers. However, the prevalent use of a molar excess of the unlabelled competitor RNA as a means of establishing specificity is insufficient, since the unlabelled RNA may also exhibit the aberrant conformation. The most important parameters in terms of specificity are the mono- and divalent cation concentrations. As illustrated in Fig. 4.5 it is feasible to obtain different cross-linking patterns by altering the concentration of Mg^{2+}, so an initial titration curve of this type is strongly recommended.

The use of photoreactive nucleoside analogues of the type listed in Table 4.1 decreases the likelihood of generating aberrant conformers during the irradiation process but the analogues may, of course, alter the conformation at least locally. Regardless of the cross-linking efficiency and the employed irradiation conditions, the status of a protein as a *bona fide* component of an examined complex ought to be verified by immunoprecipitation or by affinity-purification to tagged RNA (Section 4.3.1). If a photoreactive reagent is incorporated into a specific position by the DNA ligase approach described in Section 2.3.4, the choice of the photoreactive species is greater than that in Table 4.1, since the modified position is the initiating position in the downstream fragment.

4.4. *RNA footprinting*

The purpose of an RNA footprinting experiment is to predict the structure of the RNA and/or to identify the binding site(s) for proteins or ligands. The experimental approach for the structural analysis of RNA, or RNA-protein to complexes, utilises a number of different chemicals and ribonucleases which react with nucleotides in certain structural conformations of the RNA. The properties required of the chemicals and ribonucleases used for probing of the RNA structure are not necessarily the same as the properties required for probing of an RNA-protein complex (Fig. 4.6).

Structural analysis of RNA

Certain criteria must be fulfilled for a chemical or an enzymatic probe to be useful in a structural analysis of an RNA molecule.

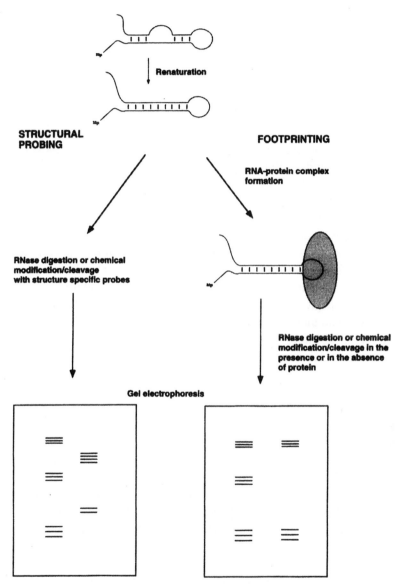

Fig. 4.6. Illustration of structural probing and footprinting of an end-labelled RNA substrate. The structural probing is performed with a single strand-specific probe (left lane) and a double-strand specific probe (right lane). The footprint is performed in the absence (left lane) or the presence of the protein (right lane) with the single strand specific probe.

1. It must be possible to determine the position of the modification.
2. It must not pertubate the structure (denature or unfold)
3. It must react in a structure-specific manner.
4. The kinetics should allow 'single-hit conditions'.

'Single-hit conditions' refer to conditions where only a single residue in the RNA molecule is modified, or the RNA is only cut at a single position. This is to ensure that the RNA molecules are structurally homogeneous since many of the reaction products after a chemical modification or cutting with an RNase, alter the structure of the RNA. Consecutive modifications (secondary modifications) might be a result of structural changes induced by the initial modification rather than reflecting the original RNA structure. In the case of small RNA molecules (<300–400 nucleotides) this is assessed by comparing the intensity of the unmodified RNA band ('mother-band') in different samples. Less than 50% of the RNA should be modified. However, for larger RNA molecules single-hit kinetics will give a modification level that is too low for further analysis. The modification level should be adjusted for the particular region of interest, but a rule of thumb is one modification per 400 nucleotides.

Protein binding sites
Protein binding sites can be studied by two different methods using RNA modifications; the footprinting technique and the modification-selection experiment. The latter is described in Section 4.6.

In the footprinting method the protein-RNA complex is formed and then modified under conditions where the complex is stable. The modification pattern of the protein-RNA complex is then compared to naked RNA modified under the same conditions. It is crucial that the reagents are reactive under physiological conditions (pH, temperature, salt) leaving the complex intact during incubation (Fig. 4.7).

The stability of the complex during the modification requires careful examination of the complex afterwards, since many of the chemicals react readily with amino acid residues, and commercially

(a)

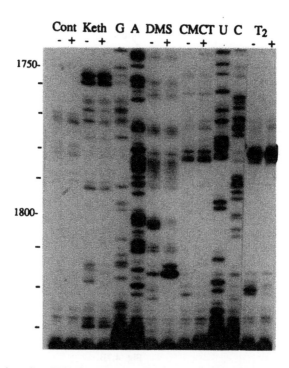

Fig. 4.7. Illustration of an RNA footprint experiment. (a) Autoradiogram showing the altered chemical and ribonuclease reactivities in the presence (+) and in the absence (−) of a binding protein. The extension is from an end-labelled primer. (b) Reactivites which are altered in the presence of the protein is indicated on a secondary structure model. Protein induced protection of RNase T1 and T2 are indicated by arrows "outside" the backbone, while arrows penetrating the backbone indicate sites protected against CVE. Bases protected against DMS (circles), kethoxal (stars) and CMCT (boxed residues) are indicated. The small vertical arrows indicate enhanced base reactivities.

available ribonucleases are frequently contaminated with proteases. Thus, it is advisable to examine the stability of the RNA-protein complex after probing, particularly if complete protection is absent. Moreover, it may also be necessary to examine the integrity of the binding protein after nuclease digests (p. 112).

(b)

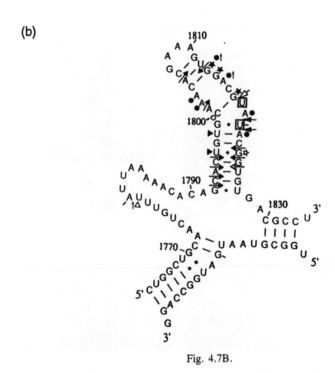

Fig. 4.7B.

4.4.1. *Enzymatic probing*

Ribonucleases are macromolecules and therefore sensitive to steric hindrance especially from the proteins in the RNA-protein complexes and from the intramolecular structures. The analysis performed with the ribonucleases described in this section provides valuable information on the gross folding of an RNA, and in many cases the mild conditions for RNase digestions are optimal for RNA-protein complexes.

Various RNases employed in different studies will not be described in this section because the optimum conditions for many of these are incompatible with proper folding of the RNA or protein-

TABLE 4.2
RNases recommended for RNA studies

	Structural specificities for ribonucleases
Single strand specific	
RNase A	$Up\downarrow A \geqslant Up\downarrow > Cp\downarrow$
RNase T1	$Gp\downarrow$
RNase T2	$Ap\downarrow > Np\downarrow$
Double strand specific	
RNase V1	$N\downarrow p$

RNA interaction and therefore cannot be recommended for probing experiments (e.g. RNase S1 and RNase U2 have a pH optimum of 4.5 which prevents or destabilises protein-RNA complex formation). RNase A has a high affinity for a pyrimidine-adenosine stretch (particularly UA), so it can therefore be difficult to obtain single-hit kinetics (except for the UA sequence). Furthermore, RNase A exhibits an endogenous helix unfolding property which makes structure assignment difficult. RNases such as RNase CL3 and α-sarcin are inhibited by Mg^{2+} which is required for the stability of many complexes and also for proper folding of RNA.

Materials
Modification buffer[a]
   70 mM HEPES-KOH, pH 7.8
   10 mM $MgCl_2$
   270 mM KCl
RNase T1 (Sigma, Pharmacia)
RNase T2 (Sigma)
RNase V1 (Pharmacia)

Procedure
1. Place 10 pmol renatured RNA or complex in 20 µl *modification buffer* on ice.
2. Add 0.01 unit *RNase T1*, **or** *0.05 unit RNase T2*, **or** *0.15 unit RNase V1*.

3. Incubate for 30 min at 0°C.
4. Terminate the reaction by adding 150 μl phenol followed by 130 μl 0.3 M NaOAc (pH 6.0).
5. Extract twice with phenol,[b] once with phenol/chloroform (1 : 1), and once with chloroform.
6. Precipitate with 2.5 × vol EtOH and wash with 80% EtOH.
7. Dissolve the pellet in 20 μl TE buffer (0.5 pmol/μl).

Note
a. The choice of buffer is less critical for ribonucleases than for chemicals. Tris-HCl is often used instead of HEPES-KOH.
b. The phenol should be saturated with 0.3 M NaOAc (pH 6.0).

Comments
RNase T1 cleaves 3′ of guanosines leaving a 3′ phosphate. It recognises the *syn* conformation of guanosines explaining the strong preference for guanosines in terminal loops.

RNase T2 cleaves at the 3′ site of the recognised base leaving a 3′ phosphate and exhibits a specificity (with a preference for adenosines) for bases located in terminal loops while the reactivity in internal loops is weaker.

Two preparations of RNases purified from cobra venom have been described. The preparation purified according to Vassilenko and co-workers, RNase CV, appears to attack double stranded regions primarily while the RNase V1 (Pharmacia) may have additional single-stranded activity. Neither RNase CV nor RNase V1 exhibit any apparent sequence specificity.

Many of the RNase preparations are contaminated with proteases, thus a test for the integrity of the protein is necessary after treatment of RNA-protein complexes.

4.4.2. Chemical probing

The advantage of chemical probes is the relatively small size of the compounds which make them sensitive to local effects, thus providing a detailed insight to the RNA structure as well as delimiting the boundaries of a protein binding site. A number of different chemicals have been used as probes but many of them are of limited use. In this section, the most useful and reliable probes are described. Taken together it is possible to probe all four nucleotides although the bases do not react evenly with the probes. As an example, the N1 position at G reacts slower than the N3 position at U's with CMCT, thus only highly reactive G's will be detected under conditions where moderate reactive U's will be modified, making CMCT a preferred U probe. Kethoxal reacts exclusively with G's at the N1 and N2 and can therefore be used as a G probe. C's are the most difficult to probe since the only available probe, DMS, react faster with N1 position at A's than the N3 on C's (Figs. 4.8 and 4.9).

Hydroxyl radical induced strand scission is used for probing of the ribose moiety. The probe is insensitive to local conformations and not useful for detailed structural predictions; however, the nonselective reactivity and small size makes hydroxyl radical probing a powerful approach for footprinting proteins on RNA. Recently methods have been developed where the hydroxyl radicals are generated locally by attaching the iron ion, which catalyses the radical formation in the Fenton reaction, to specific sites on RNA binding proteins. RNA located in the neighborhood of the tethered Fe^{2+} will then be modified selectively.[7,8]

Probing of the phosphates is informative for localisation of protein binding sites and in particular for determination of the groove that interacts with the protein. Unfortunately, phosphates are rather inert and ethylnitrosurea (ENU) which has been used as a probe reacts under very harsh conditions where most RNA-protein complexes will dissociate. However, when *in vitro* transcribed

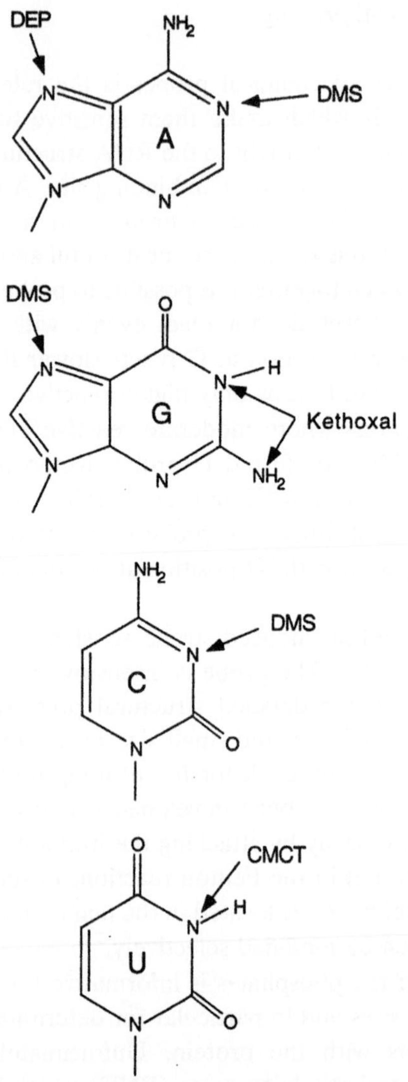

Fig. 4.8. The structure of the bases. Reactive groups are indicated by arrows.

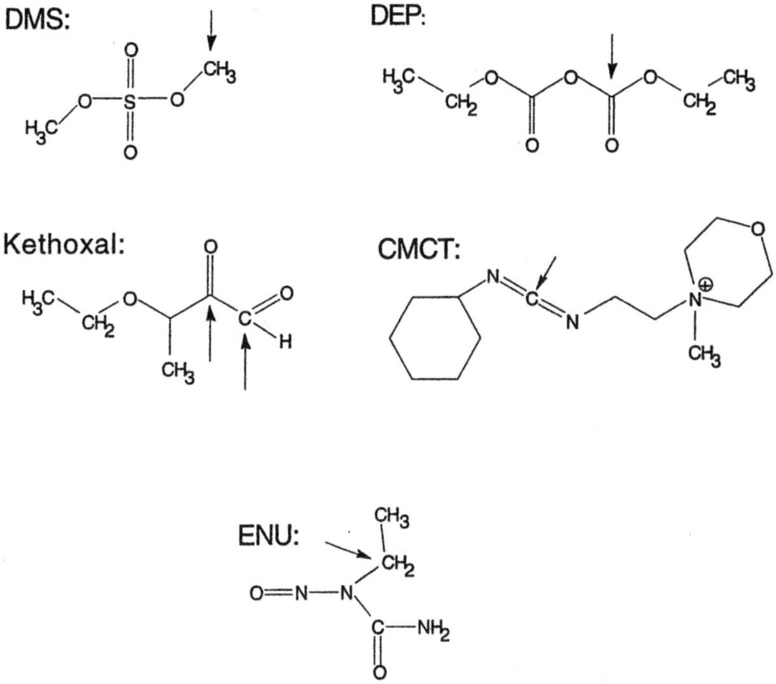

DMS:

DEP:

Kethoxal:

CMCT:

ENU:

Fig. 4.9. The chemical structure of the described probes. An arrow indicates the reactive atoms.

RNAs are studied the problems can often be circumvented by complementing the NTPs with 5% of the nucleoside 5′-O-(1-thiotriphosphate, NTPαS). The phosphorothioate is susceptible to iodine induced cleavage which is less harmful for most complexes.[9] However, as described in Section 4.6, ENU is a useful probe in damage selection experiments.

4.4.2.1. DMS (dimethylsulphate) modification for probing adenosines and cytidines
Materials
Modification buffer[a]
   70 mM HEPES-KOH, pH 7.8
   10 mM MgCl$_2$

TABLE 4.3
Chemical probes recommended for RNA modification

Structural specificities of chemical probes

Chemical	Reactive residues (atoms)	Detection method
Kethoxal	G(N1, N2)	RT
DMS	G(N7) > A(N1) > C(N3)	RT (A and C directly and G after aniline)
		EL (G after NaBH$_4$/aniline and C after hydrazine/aniline)
CMCT	U(N3) ≫ G(N1)	RT
DEP	A(N7, N6)	RT and EL (after hydrazine)
FeEDTA H$_2$O$_2$	Ribose (C4'?)	RT and EL
ENU*	Phosphate	RT and EL

RT: reverse transcriptase; EL: end-labelling.
*ENU is reactive at nonphysiological conditions and is generally not useful in footprinting experiments.

270 mM KCl
DMS termination buffer
  1.0 M Tris-acetate, pH 7.5
  1.0 M mercaptoethanol
  1.5 M sodium acetate
  0.1 mM EDTA
DMS (Sigma)

Procedure
1. Place 10 pmol renatured RNA or complex in 200 µl *modification buffer* on ice.
2. Add 1 µl of a freshly made 1 : 2 dilution of *DMS*[b] in EtOH.
3. Incubate for 5 min at 30°C or 30 min at 0°C.
4. Terminate the reaction by adding 50 µl DMS *termination buffer*.
5. Precipitate RNA by addition of 750 µl EtOH (leave on dry ice for 15 min before centrifugation).
6. Redissolve the pellet in 150 µl 0.3 M NaOAc (pH 6.0).

7. Extract once with phenol, once with phenol/chloroform (1 : 1), and once with chloroform.
8. Precipitate with EtOH and wash with 80% EtOH
9. Dissolve pellet in 20 µl TE buffer (0.5 pmol/µl).

Notes

a. HEPES has an optimal pKa (7.5) and it is sufficiently inert not to interfere with the chemical modification. The pH buffer capacity is particularly critical for DMS modifications due to the formation of sulfuric acid as a result of DMS hydrolysis.
b. All DMS waste should be inactivated in 10 M NaOH.

Comments

DMS methylates the bases by a nucleophilic attack on nitrogen atoms in the following order $G(N7) \gg A(N1) > C\,(N3)$. The reactivity of the base nitrogen is inhibited when the base is involved in a hydrogen base pairing or a metal-ion interaction. Thus the modifications at adenosines and cytidines are single-strand specific while the N7 modification of guanosines also occurs in helices at the terminal base pair. DMS will also react with base paired adenosines in *syn* conformation which occurs in G-A basepairs and A-U Hoogsteen basepairs.

The modification at N1 positions on adenosines and N3 positions on cytidines is directly monitored by the reverse transcriptase assay. However, it is important to keep in mind that conditions which give rise to a limited level of modification of adenosines and cytidines may, in addition, substantially modify guanosines at the N7 position, which may lead to secondary effects. The extent of guanosine modifications cannot be detected directly with the reverse transcriptase assay but can be assessed after reduction with sodium borohydride followed by acid-catalysed strand scission (Section 4.4.2.2).

4.4.2.2. DMS (dimethylsulphate) modification for probing guanosine

Materials

NaBH₄ stop solution
  0.6 M NaOAc
  0.6 M HOAc
  pH 4.0–5.0

Heavily methylated RNA carrier
  See Section 4.4.3.2.2

Aniline/acetic acid[a] (*1.0 M aniline solution*)
  Mix first:
    90 µl glacial acetic acid
    210 µl water
  Then add 30 µl redistilled aniline (11 M) to the water/HOAc solution, and rinse the dispensing pipette several times.

NaBH₄ 0.2 M (Merck)

Procedure

1. Modify with DMS as described in Section 4.4.2.1, but redissolve the pellet at step 9 in 10 µl 1.0 M Tris-HCl (pH 8.2) containing 8 µg *methylated carrier* RNA.[b]
2. Add 10 µl freshly prepared *0.2 M NaBH₄* and incubate for 30 min in the dark.
3. Terminate the reaction by adding 200 µl *NaBH₄ stop solution*.
4. Precipitate RNA by addition of 600 µl EtOH (leave on dry ice for 15 min before centrifugation).
5. Redissolve in 200 µl 0.3 M NaOAc (pH 6.0) and precipitate with 500 µl EtOH.
6. Wash the pellet in 80% EtOH and dry the pellet.
7. The pellet is redissolved in 20 µl aniline/acetic acid solution and incubated for 20 min at 60°C in the dark.
8. Lyophilise.
9. Redissolve in 20 µl H₂O and lyophilise.
10. Repeat step 9.
11. Redissolve in TE buffer at a concentration of 0.5 pmol/µl.

Notes

a. Prepare the aniline/acetic acid solution immediately before use. The stock solution oxidises (and turns yellow) fairly rapidly, even at −20°C. This yellow solution will catalyse the strand scission reaction but might lead to problems.

b. The presence of highly methylated RNA at the borohydride step is crucial for cleavage of the RNA, in particular when the RNA concentration is low.

4.4.2.3. DEP (diethylpyrocarbonate) modification for probing adenosines

Materials

Modification buffer
   70 mM HEPES-KOH, pH 7.8
   10 mM MgCl$_2$
   270 mM KCl
DEP[a] (Sigma)

Procedure

1. Place 10 pmol renatured RNA or complex in 200 μl *modification buffer* on ice.
2. Add 5 μl *DEP*.[b]
3. Incubate at 30°C for 10 min (or at 0°C for 30–60 min, shake every 15 min[c]).
4. Terminate the reaction by adding 20 μl 3 M NaOAc (pH 6.0) and precipitate with 600 μl EtOH (leave on dry ice for 15 min).
5. Redissolve the pellet in 150 μl 0.3 M NaOAc (pH 6.0).
6. Extract once with phenol, once with phenol/chloroform (1 : 1), and once with chloroform.
7. Precipitate with EtOH and wash with 80% EtOH.
8. Dissolve pellet in 20 μl TE buffer (0.5 pmol/μl).

Notes

a. DEP should be kept under dry conditions. In the presence of H_2O, DEP readily decompose into ethanol and carbon dioxide.
b. DEP is a potential carcinogen and all waste containing DEP should be inactivated in 10 M NaOH.
c. The solution is saturated with DEP and small drops of DEP are present in the solution. Due to the fast hydrolysis of DEP in an aqueous solution shaking is necessary to keep the solution saturated.

Comments

DEP carbethoxylates the N7 position on adenosines. The modification is inhibited by stacking of the imidazole ring, thus only adenosines in single stranded regions are modified.

DEP-modified adenosines are monitored directly in reverse transcriptase assays. The reason for the termination of the reverse transcription is not clear. It may either reflect the bulkiness of the carbethoxy group on the N7, or that the carbethoxylation of N7 causes spontaneous hydrolysis at the C8 position and therefore breakage of the imidazole ring. A third explanation is that subsequent to the N7 modification the N6 position is carbethoxylated and this modification blocks the reverse transcriptase.

DEP reacts strongly with proteins (particular histidines) so the stability of protein-RNA complexes should always be tested after modification (see Section 4.2).

Strand scission after DEP modification (for end-labelled RNA)
Materials
Aniline/acetic acid[a] (*1.0 M aniline solution*)
Mix first:
90 μl glacial acetic acid
210 μl water
Then add 30 μl redistilled aniline (11 M) to the water/HOAc solution, and rinse the dispensing pipette several times.

Procedure

1. Instead of redissolving the pellet in TE buffer (step 8; Section 4.4.2.3) add 20 μl *aniline/acetic acid* solution and incubate for 20 min at 60°C in the dark.
2. Lyophilise.
3. Redissolve in 20 μl H_2O and lyophilise.
4. Repeat step 3.
5. Redissolve in TE buffer at a concentration of 0.5 pmol/μl.

Notes

a. Prepare the aniline solution immediately before use because the solution oxidises (and turns yellow) fairly rapidly, even at −20°C. This yellow solution will catalyse the strand scission reaction but might lead to problems.

4.4.2.4. Kethoxal (3-ethoxy-2-oxo-butanal) modification for probing of guanosines

Materials

Modification buffer
   70 mM HEPES-KOH, pH 7.8
   10 mM $MgCl_2$
   270 mM KCl
TEB
   10 mM Tris-HCl, pH 7.8
   0.1 mM EDTA
   50 mM potassium borate[a]
Kethoxal precipitation buffer
   0.3 M NaOAc (pH 6.0)
   0.25 M boric acid
Kethoxal (40 mg/ml[b]) (Research Organics, Cleveland)

Procedure

1. Place 10 pmol renatured RNA or complex in 200 μl *modification buffer* on ice.
2. Add 5 μl *Kethoxal* (40 mg/ml).

3. Incubate at 30°C for 10 min (or at 0°C for 60 min).
4. Terminate the reaction by adding 20 μl *Kethoxal precipitation buffer*, and precipitate with 600 μl EtOH (leave on dry ice for 15 min).
5. Wash the pellet.
6. Redissolve the pellet in 150 μl *Kethoxal precipitation buffer*.
7. Extract once with phenol, once with phenol/chloroform (1 : 1), and once with chloroform.
8. Precipitate with EtOH and wash with 80% EtOH.
9. Dissolve pellet in 20 μl *TEB* buffer (0.5 pmol/μl).

Notes
a. Borate stabilises the tricyclic adduct formed between kethoxal and guanosines.
b. Some vendors supply Kethoxal as a lyophilised compound which is insoluble in water. The kethoxal can be dissolved in a small volume of ethanol and slowly diluted with water until a 20% ethanol solution is obtained. The final concentration should not exceed 40 mg/ml. Store at 4°C.

Comments
Kethoxal is a 1,2-dicarbonyl compound that reacts with both the N1 and N2 positions on guanosines and generate a tricyclic adduct. Kethoxal reacts with guanosines in single-stranded regions and terminal base pairs. The modified residues are monitored directly in the reverse transcriptase assay.

4.4.2.5. CMCT (1-cyclohexyl–3-(2-morpholinoethyl)-carbodiimide) for probing uridines and guanosines
Materials
Modification buffer
    70 mM HEPES-KOH, pH 7.8[a]
    10 mM MgCl$_2$
    270 mM KCl
CMCT (42 mg/ml[b]*)* (Sigma)

Procedure

1. Place 10 pmol renatured RNA or complex in 20 µl *modification buffer* on ice.
2. Add 20 µl *CMCT* (42 mg/ml).
3. Incubate at 30°C for 20 min (or at 0°C for 90 min).
4. Terminate the reaction by adding 200 µl 0.3 M NaOAc (pH 6.0) and precipitate with 600 µl EtOH (leave on dry ice for 15 min).
5. Wash the pellet with 80% EtOH.
6. Redissolve the pellet in 150 µl 0.3 M NaOAc (pH 6.0).
7. Extract once with phenol, once with phenol/chloroform (1 : 1), and once with chloroform.
8. Precipitate with EtOH and wash with 80% EtOH.
9. Dissolve pellet in 20 µl TE buffer (0.5 pmol/µl).

Notes

a. The reactivity of CMCT depends strongly on the pH, that should not be lower than 7.8. Increasing the pH will increase the reactivity.
b. CMCT is stored as a 42 mg/ml stock solution in the appropriate buffer, i.e. 70 mM HEPES-KOH, pH 7.2, 10 mM $MgCl_2$, 270 mM KCl at 4°C.

Comments

CMCT modifies uridines at N3 in single-stranded regions and with lower reactivity guanosines at the N1 position, and the modifications are directly monitored in the reverse transcriptase assay.

4.4.2.6. Hydroxyl radical modification for probing ribose moieties
Materials
Modification buffer
   70 mM HEPES-KOH pH 7.8
   10 mM $MgCl_2$
   270 mM KCl

Hydroxyl radical solution[a] (per reaction)
   2.5 μl 50 mM $Fe(NH_4)_2(SO_4)_2$ $6H_2O$
   2.5 μl 100 mM EDTA, pH 8.0
   2.5 μl 250 mM ascorbate
   2.5 μl 2.5% H_2O_2

Procedure
1. Place 10 pmol renatured RNA or complex in 25 μl *modification buffer* on ice.
2. Add 10 μl freshly made *hydroxyl radical solution*.
3. Incubate at 0°C for 4 min.
4. Terminate the reaction by adding 200 μl 0.3 M NaOAc (pH 6.0) and precipitate with 600 μl EtOH (leave on dry ice for 15 min).
5. Wash the pellet with 80% EtOH.
6. Redissolve the pellet in 150 μl *0.3 M NaOAc* (pH 6.0).
7. Extract once with phenol, once with phenol/chloroform (1 : 1), and once with chloroform.
8. Precipitate with EtOH and wash with 80% EtOH.
9. Dissolve pellet in 20 μl TE (0.5 pmol/μl).

Note
a. The hydroxyl radical, Fe^{2+} and H_2O_2 solutions should always be made immediately before use.

Comments
The mechanism for hydroxyl-induced strand cleavage is not completely understood. The ribose carbons C1′ and C4′ are the predominant sites of attack, i.e. the hydroxyl radical absorbs the hydrogen leaving an unstable C1′ or C4′ radical. The C1′ and C4′ positions are accessible both in nucleotides involved in base-pairing and in nucleotides located in single-stranded regions. Thus hydroxyl radical probing will not distinguish between different secondary structures but nucleotides located in the interior of large RNA molecules or involved in tertiary structures exhibit reduced reactivity.

The insensitivity to different RNA structures combined with the small size of the hydroxyl radical makes it a powerful probe for footprinting proteins on RNA. In particular protein interactions with minor groove, since both C1' and C4' are located in the minor groove of helices. The stability of protein-RNA complexes should be tested after modification (see Section 4.2) since hydroxyl radicals also react with proteins.

4.4.2.7. Iodine scission of phosphorothioate containing RNA for probing phosphates

The inertness of the phosphate groups to chemical probes under physiological conditions can be circumvented by co-transcriptional substitution of the phosphate with the more reactive phosphorothioate. The RNA is generated as described in Section 2.3.2 by *in vitro* transcription including 5% of the nucleoside 5'-O-(1-thiotriphosphates) (NTPαS) in the reaction mixture. The subsequent iodine probing results in cleavage of the backbone that can be monitored by either end-labelling or reverse transcriptase analysis.

Material

Modification buffer
    70 mM HEPES-KOH, pH 7.8
    10 mM MgCl$_2$
    270 mM KCl
Iodine (2 mM, Sigma)

Procedure

1. Place 10 pmol renatured phosphorothioate-containing RNA or complex in 20 µl *modification buffer* on ice.
2. Add 2 µl of a 2 mM *iodine* solution.[a]
3. Incubate for 1 min.
4. Terminate the reaction by adding 200 µl 0.3 M NaOAc (pH 6.0) and precipitate with 600 µl EtOH (leave on dry ice for 15 min).
5. Wash the pellet with 80% EtOH.
6. Redissolve the pellet in 150 µl 0.3 M NaOAc (pH 6.0).

7. Extract once with phenol, once with phenol/chloroform (1 : 1), and once with chloroform.
8. Precipitate with EtOH and wash with 80% EtOH.
9. Dissolve pellet in 10 μl TE buffer (0.5 pmol/μl) or in loading buffer if the RNA is end-labelled.

Note
a. The iodine concentration may have to be optimised for different experiments.

Comments
In structural studies, a low concentration of iodine is advisable in order to obtain a structure-dependent modification pattern. In footprinting of proteins on RNA the optimal conditions should permit reaction with all the free phophorothioates (i.e. those not interacting with the protein). Since iodine also reacts strongly with proteins the quantitative strand scission may be limited by the dissociation of the protein-RNA complex.

4.4.3. Identification of modified residues

There are two different methods to identify modified residues and ribonuclease scissions in RNA molecules; the reverse transcriptase method or the end-labelling method. The choice of method depends both on the length of the studied RNA and the method of probing (as discussed in Sections 4.4.1 and 4.4.2). The reverse transcription method uses extension of a primer and therefore is independent of the length of the RNA, while the end-labelling method is restricted to analysis of small RNA molecules ($n < 300$). The latter method requires scission of the RNA.

4.4.3.1. Reverse transcription assay
The basis of a reverse transcriptase assay is to extend a deoxynucleotide primer, annealed to the RNA, by reverse transcriptase.

Modified bases in the RNA will stop or pause the reverse transcriptase 3' to the modified base. Breaks in the RNA induced by ribonucleases or chemicals will also terminate the polymerase. The electrophoretic analysis of the reverse transcript reveals the stops as bands on an autoradiograph. The position of the modification or scission can be determined by co-electrophoresis of a dideoxy sequence reaction, and the intensity of the band will reflect the reactivity of the base (Figs 4.7 and 4.10).

The radioactive label can be incorporated into the reverse transcript either (a) by the use of a 5' end-labelled primer, or (b) by incorporation of α-labelled deoxynucleoside triphosphates during reverse transcription. The latter method is used in combination with an unlabelled primer. The bands generated by method (a) originates from fragments of identical specific activity (one radioactive atom per fragment), and RNA-self priming will not produce any artefact bands. However, the band intensities obtained from longer fragments will not reflect the fraction of modified RNA molecules, but a smaller fraction, because the reverse transcriptase will terminate prematurely at nicks in the RNA, by certain RNA structures, or at endogenous modified bases. These stops will also appear in the control lanes (untreated samples). Control bands originating from nicks (breakdown) of the RNA are most prevalent at low primer : RNA ratios where intramolecular RNA–RNA interactions in full-length molecules compete with the primer annealing. The low primer : RNA ratio is easily alleviated in the other method (b) where an unlabelled primer is used for the extension. Moreover method (b) compensates for the premature termination of the reverse transcriptase by incorporating additional radioactive residues in the longer transcripts. However, this procedure may generate artefact bands due to RNA self-priming. Therefore, it is important to keep in mind that comparisons of band intensities in the same lane after a reverse transcriptase extension must be exercised with great caution. Comparisons should be restricted to bands of identical size. Since the detection of ligand or protein interactions

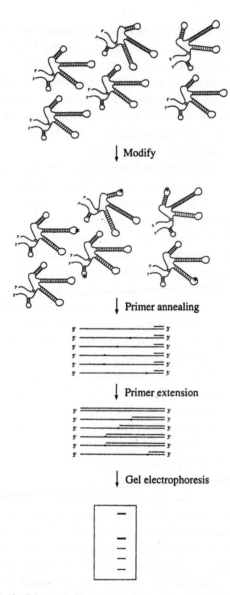

Fig. 4.10. Schematic illustration of a reverse transcription assay.

is based on comparisons between reverse transcripts of equal length these considerations do not normally impose limitations on the analysis of ligand binding sites.

The identification of the modified residue is established when the sample is co-electrophoresed with a sequence (Section 4.4.3.1.3). Since the reverse transcriptase stops or pauses 3′ of the modified band, the dideoxy sequencing bands are displaced by *one base* relative to the modified base, i.e. if a residue is modified the corresponding band will appear as one nucleotide shorter than the base will appear in the sequence lane.

4.4.3.1.1. Extension from an end-labelled primer
Materials
10 × anneal buffer
    100 mM Tris-HCl, pH 7.4
    400 mM KCl
    5 mM EDTA
25 × RT buffer
    1.25 M Tris-HCl, pH 8.0
    250 mM MgCl$_2$
    50 mM DTT
Maxam-Gilbert load buffer
    80% deionised formamide
    10 mM NaOH
    1 mM EDTA
    0.02% xylene cyanol
    0.02% bromophenol blue
dNTP 2.5 mM
AMV reverse transcriptase
5′-[^{32}P] primer (~3000 Ci/mmol)

Procedure
Annealing
1. Mix 0.5 pmol RNA and 0.5 pmol 5′-[^{32}P] primer in 6 µl *1 × anneal buffer*.

2. Heat to 95°C for 30 s, transfer to 50°C and leave for 20 min.
3. Spin down evaporated drops.

Extension

4. Add 4 µl extension mixture (made of 0.4 µl *25 × RT buffer*, 0.8 µl 2.5 mM of each *dNTPs*, 2.8 µl H_2O and 1 unit *AMV reverse transcriptase*[a]) to the annealed primer/template solution.
5. Incubate for 30 min at 37–42°C.
6. Stop the reaction by addition of 40 µl 0.3 M NaOAc (pH 6.0) and precipitate with 125 µl EtOH.
7. Dissolve the washed and dried pellet in *Maxam-Gilbert load buffer*.[b]
8. Heat the samples to 95°C for 3 min and chill on ice.
9. Load 1–3 µl on a sequencing gel.

Notes

a. Excess of reverse transcriptase can cause read-through or extension by one nucleotide at certain modifications or addition of an additional nucleotide after strand breakage.
b Maxam-Gilbert load buffer is used to hydrolyse the RNA before the electrophoretic analysis. Large RNA fragments can interfere with the mobility of the reverse transcripts.

4.4.3.1.2. Extension in the presence of radioactive dATP

Materials

10 × anneal buffer
　100 mM Tris-HCl, pH 7.4
　400 mM KCl
　5 mM EDTA
25 × RT buffer
　1.25 M Tris-HCl, pH 8.0
　250 mM $MgCl_2$
　50 mM DTT
Maxam-Gilbert load buffer
　80% deionised formamide
　10 mM NaOH

1 mM EDTA

0.02% xylene cyanol

0.02% bromophenol blue

dGTP, dCTP, dTTP mix 0.75 mM each

dATP 0.75 mM

AMV reverse transcriptase

α-[^{35}S]*dATP* (1000 Ci/mmol; 10 mCi/ml), **or** α-[^{32}P]*dATP* (3000 Ci/mmol; 10 mCi/ml)

Procedure

Annealing

1. Mix 0.5 pmol RNA and 0.5 pmol primer in 6 µl 1 × *anneal buffer*.
2. Heat to 95°C for 30 sec, transfer to 50°C and leave for 20 min.
3. Spin down evaporated drops.

Extension

4. Mix 3 µl of the annealed primer/template solution with 2 µl extension mixture made of 0.2 µl *25 × RT buffer*, 0.2 µl α-[^{35}S]*dATP* (1000 Ci/mmol; 10 mCi/ml), **or** *0.2 µl α-[^{32}P]dATP* (3000 Ci/mmol; 10 mCi/ml), 1.6 µl 0.75 mM *dGTP, dCTP, dTTP mix* and 1 unit *AMV reverse transcriptase*.[a]
5. Incubate at 37–42°C for 15 min.
6. Add 2 µl 0.75 mM dATP and continue the incubation for 15 min.
7. Stop the reaction by addition of 5 µl *Maxam-Gilbert load buffer*.[b]
8. Denature the samples by heating to 95°C for 3 min and chill on ice.
9. Load 1–3 µl on a sequencing gel.

Notes

a. Excess of reverse transcriptase can cause read-through or extension by one nucleotide at certain modifications or addition of an additional nucleotide after strand breakage.

b Maxam-Gilbert load buffer is used to hydrolyse the RNA before the electrophoretic analysis. Large RNA fragments can interfere with the mobility of the reverse transcripts.

4.4.3.1.3. Sequencing by reverse transcriptase method

Materials

10 × anneal buffer
　100 mM Tris-HCl, pH 7.4
　400 mM KCl
　5 mM EDTA
25 × RT buffer
　1.25 M Tris-HCl, pH 8.0
　250 mM MgCl$_2$
　50 mM DTT
Maxam-Gilbert loading buffer
　80% deionised formamide
　10 mM NaOH
　1 mM EDTA
　0.02% xylene cyanol
　0.02% bromophenol blue
ddATP 2.0 mM
ddTTP 2.0 mM
ddCTP 0.66 mM
ddGTP 1.33 mM
dNTPs 2.5 mM
AMV reverse transcriptase
End-labelled primer

Procedure

Annealing
　1. Mix
　　　　5 pmol RNA
　　　　2.5 pmol 5'-[^{32}P] primer (~3000 Ci/mmol)
　　　　2.5 μl *10 × anneal buffer*
　　　　H$_2$O to a total of 25 μl

2. Heat at 95°C for 2 min and transfer to 50°C and leave for 20 min.
3. Dispense the annealing mixture into four 6-μl aliquots at room temperature.

Extension

4. Add 1 μl of the following four dideoxynucleoside triphosphate solutions to each of the 6-μl aliquots:

A reaction	*2.0 mM ddATP*
T reaction	*2.0 mM ddTTP*
C reaction	*0.66 mM ddCTP*
G reaction	*1.33 mM ddGTP*

and mix.[a]

5. Start extension by adding 3 μl of a mix made up from

0.4 μl *25 × extension buffer*
0.8 μl *2.5 mM dNTPs*
1.8 μl H_2O
1 unit *AMV reverse transcriptase*.

6. Incubate at 37°C for 30 min (raise temperature to 42°C if premature termination is a problem).
7. Stop extension by the addition of 10 μl *0.5 M* NaOAc (pH 6.0) and 50 μl EtOH.
8. Pellet, wash and dry.
9. Redissolve in *Maxam-Gilbert loading buffer*.
10. Denature at 95°C for 3 min prior to loading.

Note

a. The ddNTP concentrations are optimised for sequencing 200–300 nucleotides. The ddNTP concentration can be raised 2-fold for short RNA molecules or lowered 2-fold for longer readings.

4.4.3.2. End-labelling method

This method is useful for small RNAs and complements the primer extension method at the 3′ end of the RNA where information is lost due to primer annealing. The method requires RNA with at least one homogeneous end. Furthermore, strand scission is re-

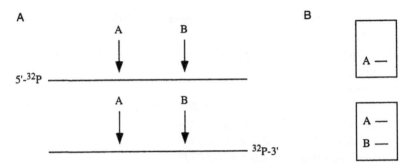

Fig. 4.11. Detection of secondary cuts by parallel labelling of the 5'-end and the 3'-end. Cut at B is secondary to A (i.e. cutting at B requires a primary cut at A).

quired for detection of the modifications. Obviously, this will not impose any limitations on the use of ribonucleases, but chemical reagents are limited to those where the modification can be converted quantitatively into a strand scission. One advantage of using 5' and 3' end-labelled RNA in parallel is that it allows distinction between primary and secondary cuts, which frequently occurs in probing experiments with ribonucleases (see Fig. 4.11).

Care should be taken in the assignment of the modified residues. This is most conveniently done by co-electrophoresis of a sequencing track. However, as illustrated in Fig. 4.12 the termini generated by the different modifications and sequencing procedures vary so the fragments may not co-migrate.

The reactions used in the chemical sequencing method employ acidic aniline to induce the strand scission leaving a modified ribose at the 3' terminus and the phosphate at the 5 terminus. In contrast, enzymatic sequencing leaves the phosphate at the 3' end and a hydroxyl group at the 5' end. The phosphate at the 3' terminus can either be in a 2', 3' cyclic form or in the normal 3' position. Since the cyclic form has one negative charge less than the 3' form it can result in double bands when the 5'-labelled RNA is sequenced. However, the effect is most pronounced for short fragments while the separation is often invisible for larger fragments.

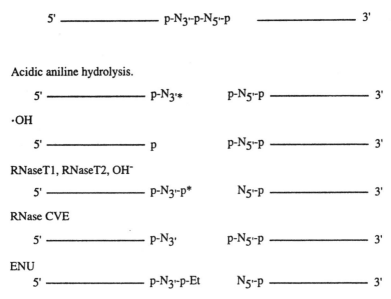

Fig. 4.12. RNA termini generated after various strand scission methods. Asterisks indicate chemical heterogeneities.

The shift in electrophoretic mobility between fragments produced by the different chemistries is often less than the distance between two succeeding nucleotides (e.g. hydroxyl radical modified 5'-labelled RNA will run slightly faster than the corresponding sequence obtained by the chemical method and exactly one nucleotide faster when compared with an enzymatic sequence). However, if a 3'-labelled RNA is analysed the fragment modified by the hydroxyl radical will co-migrate with the corresponding chemical sequence but slightly faster than the enzymatic sequence. Therefore, it is recommended to select a sequencing method which generates the same termini as the probe.

4.4.3.2.1. Enzymatic sequencing of end-labelled RNA
The enzymatic sequencing procedure often referred to as the Donis-Keller method[10] is based on a partial digestion of end-labelled

RNA with nucleotide-specific RNases. The digests produce G, A, A + U and U + C lanes. Three reactions are performed under denaturing conditions in the presence of urea that ensures a uniform band intensity independent of the secondary structure. The U + C reaction is performed in the absence of denaturing agents and consequently the band intensities depend on the secondary structure.

In addition to the sequence lanes, it is helpful to co-electrophorese an alkaline hydrolysis ladder to resolve ambiguities.

Materials
End-labelled RNA (see Section 2.3.3)
G buffer
  20 mM Na citrate, pH 5.0
  1 mM EDTA
  7 M Urea
  0.025% xylene cyanol
  0.025% bromphenol blue
A buffer
  20 mM Na citrate, pH 3.5
  1 mM EDTA
  7 M Urea
  0.025% xylene cyanol
  0.025% bromphenol blue
BC buffer
  20 mM Na citrate, pH 5.0
  1 mM EDTA
  0.025% xylene cyanol
  0.025% bromphenol blue
2 × alkaline hydrolysis buffer
  100 mM $NaHCO_3/Na_2CO_3$ pH 9.0
  2 mM EDTA
  0.5 μg/μl tRNA carrier
Urea load buffer
  8 M urea

20 mM Tris-HCl, pH 7.4
1 mM EDTA
0.05% xylene cyanol
0.05% bromphenol blue
tRNA carrier (5 μg/μl in H_2O)
RNase T1 (Pharmacia)
RNase U2 (Pharmacia)
RNase PhyM (Pharmacia)
RNase B. cereus (Pharmacia)

Procedure
1. Add the *end-labelled RNA* and 1 μl *tRNA carrier* to four tubes.
2. Dry the samples down (in a speed-vac)
3. Dissolve the pellets; G reaction: 10 μl *G buffer*
                     A reaction: 10 μl *A buffer*
                A + U reaction: 10 μl *G buffer*
                U + C reaction: 10 μl *BC buffer*
4. Heat the samples to 50°C for 5 min.
5. Chill quickly
6. Add for the G reaction: 1 μl *RNase T1* (0.05 U/μl)
                A reaction: 1 μl *RNase U2* (0.005 U/μl)
                A + U
                  reaction: 1 μl *RNase PhyM* (1 U/μl)
                U + C
                  reaction: 1 μl *RNase B. cereus* (3 U/μl)
7. Incubate at 50°C for 15 min.
8. Add 5 μl *urea load buffer* to the U + C reaction.
9. Freeze and keep frozen until ready for loading on a sequencing gel.

Alkaline hydrolysis
1. Mix an equal volume of end-labelled RNA and *2 × alkaline hydrolysis buffer*.
2. Incubate the tube for 10 min in boiling water.[a]
3. Add 1 × vol. *urea load buffer*.

Note

a. The alkaline hydrolysis ladder can be made by mixing samples treated for different time periods (2–20 min). Short readings require longer heat treatment.

Comments

An alkaline hydrolysis ladder and two of the sequence reactions are sufficient for an unambiguous sequence assignment in RNA of known sequence.

4.4.3.2.2. Chemical sequencing of end-labelled RNA. The chemical sequencing procedure is basically as described by Peattie.[11] The principle is to modify the nucleotide in a base-specific manner that makes the backbone susceptible to strand scission by the subsequent acidic aniline treatment.

Materials
G reaction:
G buffer
   50 mM sodium cacodylate adjusted with HCl, pH 5.5
   1 mM EDTA
DMS termination (pptn) solution
   1.0 M Tris-acetate, pH 7.5
   1.0 M 2-mercaptoethanol
   1.5 M sodium acetate
   0.1 mM EDTA
NaBH₄ stop solution
   0.6 M NaOAc
   0.6 M HOAc
   pH 4.0–5.0
Heavily methylated RNA carrier[12]
1. Dissolve 1 mg RNA carrier in 300 µl *G buffer*.
2. Add 5 µl *DMS* and incubate at 90°C for 5 min.
3. Add 75 µl *G precipitation solution* and 900 µl EtOH.

4. Precipitate.

5. Reprecipitate the pellet from 300 μl 0.3 M NaOAc (pH 6.0).

6. Dissolve the pellet in 1 M Tris-HCl (pH 8.2) at 8 μg/ml.

DMS (Sigma)

NaBH₄ (0.2 M)[a]

A > G Reaction:

A,G buffer

  50 mM NaOAc, pH 4.5

  1 mM EDTA

A,G precipitation (pptn) solution

  1.5 M NaOAc, pH 6.0

DEP

C > U Reaction:

Hz/NaCl

  Dissolve NaCl, oven-dried, to 3.0 M in hydrazine.

  Keep anhydrous. Cool to 0°C.[b]

C,U precipitation(pptn) solution

  80% ethanol/20% (vol/vol) water. Store and use at −20°C.

U Reaction:

Hz/H₂O

  50% hydrazine/50% water (vol/vol), cool to 0°C.

U precipitation(pptn) solution

  0.3 M NaOAc, pH 6.0

  0.1 mM EDTA

Aniline reaction:

Aniline/acetic acid (1.0 *M aniline solution*)[c]

  Mix first

  90 μl glacial acetic acid

  210 μl water.

  Then add 30 μl redistilled aniline (11 M) to the water/HOAc solution, and rinse the dispensing pipette several times.

Urea load buffer

  8 M urea

  20 mM Tris-HCl, pH 7.4

  1 mM EDTA

0.05% xylene cyanol
0.05% bromphenol blue
tRNA carrier (5 mg/ml)

Procedures

G reaction

1. Dilute end-labelled RNA and 2 μl *tRNA carrier* in 300 μl *G buffer* (chill to 0°C).
2. Add 1 μl *DMS* and treat for 1 min at 90°C.
3. Place the tube on ice; add 75 μl *DMS termination* (*pptn*) *solution* (0°C) and precipitate with 900 μl EtOH (leave on dry ice for 15 min).
4. Redissolve the pellet in 200 μl 0.25 M NaOAc (pH 6.0) and precipitate with 500 μl EtOH.
5. Wash and dry the pellet.
6. Redissolve the pellet in 10 μl 1.0 M Tris-HCl (pH 8.2) containing 8 μg heavily methylated carrier RNA.[d]
7. Add 10 μl freshly made *NaBH₄* (0.2 M) and incubate at 0°C for 30 min in the dark.
8. Add 200 μl *NaBH₄ stop solution* and precipitate with 500 μl EtOH (leave on dry ice for 15 min).
9. Wash and dry the pellet.
10. The dry pellet is ready for the acidic aniline induced strand scission (see below).

A > G reaction

1. Dilute end-labelled RNA and 2 μl *carrier tRNA* in 200 μl *A, G buffer* (chill to 0°C).
2. Add 1 μl *DEP* and treat for 5 min at 90°C.
3. Place the tube on ice, add 50 μl *A, G precipitation* (*pptn*) *solution* (0°C) and precipitate with 750 μl EtOH (leave on dry ice for 15 min).
4. Redissolve the pellet in 200 μl 0.25 M NaOAc (pH 6.0) and precipitate with 500 μl EtOH.
5. Wash and dry the pellet.

6. The dry pellet is ready for the acidic aniline induced strand scission (see below).

C > U reaction

1. Mix the end labelled RNA and 2 µl *carrier tRNA*.
2. Lyophilise (use speed vac).
3. Dissolve the pellet in 10 µl *Hz/NaCl* (0°C) and incubate 20 min at 0°C.[e]
4. Add 500 µl *C, U precipitation solution* (0°C) and precipitate (leave on dry ice for 15 min).
5. Redissolve the pellet in 200 µl 0.25 M NaOAc (pH 6.0) and precipitate with 500 µl EtOH.
6. Wash and dry the pellet.
7. The dry pellet is ready for the acidic aniline induced strand scission (see below).

U reaction

1. Mix the end-labelled RNA and 2 µl *carrier tRNA*.
2. Lyophilise (use speed vac).
3. Dissolve the pellet in 10 µl *Hz/H$_2$O* (0°C) *and incubate 5 min at 0°C*.
4. *Add* 200 *µl U precipitation (pptn) solution* (0°C) and precipitate with 750 µl EtOH (leave on dry ice for 15 min).[f]
5. Redissolve the pellet in 200 µl 0.25 M NaOAc (pH 6.0) and precipitate with 500 µl EtOH.
6. Wash and dry the pellet.
7. The dry pellet is ready for the acidic aniline induced strand scission (see below).

Acidic aniline strand scission

1. Dissolve the pellet in 20 µl *aniline/acetic acid*.
2. Incubate at 60°C for 20 min in the dark.
3. Lyophilise.
4. Redissolve in 20 µl H$_2$O.
5. Lyophilise.
6. Repeat steps 4 and 5.
7. Redissolve in *urea load buffer*. Denature the RNA at 90°C for 30 s and then quick chill on ice before loading on gel.

Overview of chemical sequencing

G	A > G	C > U	U
300 μl G buffer	200 A, G buffer	2 μl carrier tRNA	2 μl carrier tRNA
2 μl carrier tRNA	2 μl carrier tRNA	5 μl RNA-^{32}p('3)	5 μl RNA-^{32}p('3)
5 μl RNA-^{32}p('3)	5 μl RNA-^{32}p('3)	*Lyophilise*	*Lyophilise*
Chill to 0°C	*Chill to 0°C*	10 μl Hz/NaCl, 0°C	10 μl Hz/H$_2$O, 0°C
0.5 μl DMS	1 μl DEP	*Mix, spin*	*Mix, spin*
90°C, 1 min	90°C, 5 min	0°C, 20 min	0°C, 5 min
0°C	0°C	–	–
75 μl G pptn, 0°C	50 μl A, G pptn, 0°C	–	200 μl U pptn,
900 μl EtOH, 0°C	750 μl EtOH, 0°C	500 μl C, U pptn,	750 EtOH, 0°C
–70°C, 15 min	–70°C, 15 min	–70°C, 15 min	–70°C, 15 min
Pellet	*Pellet*	*Pellet*	*Pellet*
200 μl 0.25 M Na	200 μl 0.25 M Na	200 μl 0.25 M Na	200 μl 0.25 M Na
acetate + 500 μl	acetate + 500 μl	acetate + 500 μl	acetate + 500 μl
EtOH, Pellet	EtOH, Pellet	EtOH, Pellet	EtOH, Pellet
EtOH.wash, dry	*EtOH wash, dry*	*EtOH wash, dry*	*EtOH wash, dry*
10 μl 1.0 M Tris-HCl			
pH 8.2 containing			
8 μg of methy-lated carrier RNA	–	–	–
10μl 0.2 M NaBH$_4$	–	–	–
0°C, 30 min/dark	–	–	–
200μl NaBH$_4$, stop	–	–	–
600μl EtOH	–	–	–
Pellet	–	–	–
200μl 0.25 M Na			
acetate + 500 μl			
EtOH, Pellet	–	–	–
EtOH wash, dry	–	–	–
20 μl Aniline/ HOAc	20 μl Aniline/ HOAc	20 μl Aniline/ HOAc	20 μl Aniline/ HOAc
60°C, 20 min/dark	*60°C, 20 min/dark*	*60°C, 20 min/dark*	*60°C, 20 min/dark*
Lyophilise	*Lyophilise*	*Lyophilise*	*Lyophilise*
20 μl water	20 μl water	20 μl water	20 μl water
Lyophilise	*Lyophilise*	*Lyophilise*	*Lyophilise*
20 μl water	20 μl water	20 μl water	20 μl water
Lyophilise	*Lyophilise*	*Lyophilise*	*Lyophilise*
5 μl buffer/dyes	5 μl buffer/dyes	5 μl buffer/dyes	5 μl buffer/dyes
90°C, 30 sec, 0°C	*90°C, 30 sec, 0°C*	*90°C, 30 sec, 0°C*	*90°C, 30 sec, 0°C*
Load on gel	*Load on gel*	*Load on gel*	*Load on gel*

Notes

a. The $NaBH_4$(0.2 M) should be made immediately before use.

b. The Hz/NaCl solution should be fresh. Cool to 0°C and use within 10–20 min. The salt can crystallise out if the solution is left on ice too long.

c. Prepare an aniline solution immediately before use because the solution oxidises (and turns yellow) fairly rapidly, even at −20°C. (This yellow solution will catalyse the strand scission reaction but might lead to problems.)

d. The presence of highly methylated RNA at the borohydride step is crucial for quantitative cleavage of low concentrations of RNA.

e. If water is present during this reaction, the C cleavage decreases significantly and the U cleavage predominates.

f. Use a new pipette tip to remove each supernatant from the ethanol precipitations; otherwise, hydrazine is carried over from each precipitation and significant blurriness occurs on the sequencing gel.

4.5. Protein footprinting

Protein footprinting is a useful method for mapping amino acids involved in intra- and intermolecular interactions. The method is similar to standard nucleic acid footprinting, except that proteins are probed instead of nucleic acids, and proteinases are used instead of endonucleases. The peptide cleavage products are subsequently resolved in SDS polyacrylamide gels and visualised by autoradiography (Fig. 4.13). By this method it is possible to probe the accessibility of individual amino acids in the context of the protein structure (surface prediction) and intermolecular interactions (ligand footprinting). A decrease in accessibility when probing the protein in the presence of a particular ligand, may be interpreted as sterical

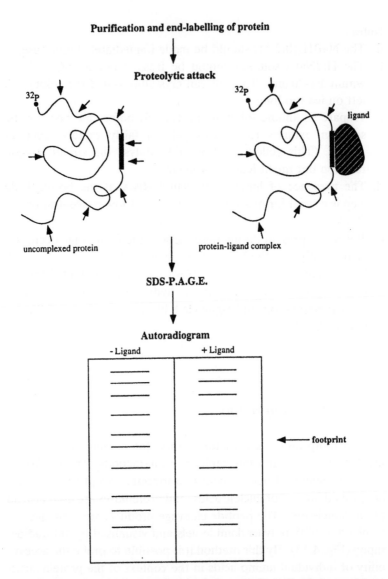

Purification and end-labelling of protein

Proteolytic attack

uncomplexed protein   protein-ligand complex

SDS-P.A.G.E.

Autoradiogram

- Ligand   + Ligand

footprint

Fig. 4.13. Flow chart of the protein footprinting technique. Purified recombinant protein, which has been radioactively labelled at the N- or C-terminal, is cleaved by proteinases or chemicals with single-hit kinetics. The substrate is kept at native conditions in uncomplexed and complexed form. The cleavage products are analysed in high resolution SDS gels.

hindrance by the ligand, thereby identifying the site of interaction. In addition, structural changes in the protein upon binding of the ligand may inhibit or enhance cleavages in regions outside the binding region.

Recently, a number of different protein footprinting procedures utilising specific detection of C- or N-terminal fragments have been published. They differ mainly in the method used for preparation of the protein and for detection of the cleavage products. Some procedures use native protein which is visualised by end-terminal specific antibodies in a Western blot.[13] The advantage of using native protein in this approach must be weighted against the cumbersome preparation of end-specific antibodies and the inability to resolve the terminal regions of the protein. Another published approach involves a purification step of end-terminal fragments by affinity purification prior to SDS-gel analysis[14] or detection by mass spectroscopy.[15] The procedure outlined below utilises radio-labelled protein which is achieved by specific labelling of a heart muscle kinase site fused to the N- or C-terminal end of the protein.[16–18] The general usefulness of this method has been increased by the commercial availability of expression vectors containing a heart muscle kinase site (HMK), and vectors designed especially for protein footprinting have recently been developed[25] (see Table 4.4). The advantage of these constructs is the opposite positioning of the purification tag and the HMK site. Thus, only the purified full-length protein will be labelled, which reduces background problems in protein footprinting experiments significantly. The GST-tag may subsequently be removed by site specific cleavage by thrombin endoproteolysis. A variant of these expression plasmids with His-tags have also been developed.[19] Since certain proteins are inactivated by end-modifications it is important to test the functional integrity of the protein that is analysed.

The interacting ligand producing the footprint may be DNA,[13,20] RNA (Fig. 4.14[16]), or protein.[17,18,21,22] The procedure below focuses on protein-RNA interactions but it is applicable for several other types of ligands.

TABLE 4.4

Bacterial expression vectors developed for protein footprinting

Name	Features	Promoter	Bacterial host
pGEX-GTH[25]	N-terminal GST-tag C-terminal HMK-site Thrombin cleavage site	*tac*	*BL21, XL-1 BLUE*
pET-HTG[25]	C-terminal GST-tag N-terminal HMK-site Thrombin cleavage site	*T 7/laco*	*BL21(DE3)*
pGEX-GEH[25]	N-terminal GST-tag C-terminal HMK-site Enterokinase cleavage site	*tac*	*BL21, XL-1 BLUE*
pET-HEG[25]	C-terminal GST-tag N-terminal HMK-site Enterokinase cleavage site	*T 7/laco*	*BL21(DE3)*
pET-His-H[19]	N-terminal His-tag C-terminal HMK-site	*T 7/laco*	*BL21(DE3)*
pET-H-His[19]	N-terminal HMK-site C-terminal His-tag	*T 7/laco*	*BL21(DE3)*

A common feature of all expression plasmids is the opposite positions of the purification- and kinase tags. See Section 3.2 for expression procedures.

Both endoproteinases and chemicals can be used to cleave the protein, provided they are active at conditions optimal for complex formation. Table 4.5 provides a list of 13 commercially available proteinases which are useful for protein footprinting. The majority of chemicals reactive towards proteins suffers from the drawback of being reactive towards nucleic acids as well. Hydroxyl radicals produced from H_2O_2 in the vicinity of Fe^{2+} ions are, however, useful for protein surface predictions and for mapping DNA-[13] and RNA binding sites (Fig. 4.14[23]) on proteins.

4.5.1. Protein footprinting using proteinases

A large collection of endoproteinases is available and they are crudely divided into two main groups, the specific and unspecific proteinases, depending on their level of specificity (summarised in Table 4.5). The specific proteinases, in general, yield few cleavages

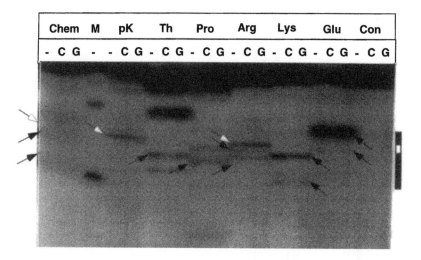

Fig. 4.14. Autoradiogram of a protein footprinting gel. Protein footprinting of poly(rC) (RNA) on the N-terminal KH domain of PCBP1 (protein) using hydroxyl radical cleavage (Chem), proteinases as indicated and no reagent (Con). The protein binds specifically to poly(rC) but not to poly(rG). Cleavages that are either inhibited or enhanced by poly(rC) are indicated by filled and open arrows, respectively. See Leffers et al. for details.[23]

in the protein and provide a good protein sequence marker. The group of unspecific proteinases, which have less stringent sequence requirements for cleavage, encompasses many members, some of which are closely related to each other in terms of their cleavage preferences. We find that a useful set of proteinases exhibiting diverse specificities include: Arg-C, Glu-C, Lys-C and Asp-N, trypsin, bromelain, chymotrypsin, elastase, pronase, proteinase K, thermolysin, subtilisin Carlsbergensis. Specificities and suggested concentrations for these proteinases are summarised in Table 4.5. Since the reactivity of proteinases towards different proteins varies considerably, the choice of reagent and conditions given in the procedure below is only a guideline and should be optimised for individual proteins.

TABLE 4.5
Proteinases used in protein footprinting

Endoproteinase	Type	Primary specificity	Comments	Suggested concentration
Arg-C	Serine	$-Arg \downarrow p_1$	retains 90% activity after 1 hr at 25 °C in 0.1% SDS	5–50 pg/µl
Asp-N	Unknown	$-p_{-1} \downarrow$ Asp	Active in 0.01% SDS, freeze/thawing not recommended	5–50 pg/µl
Bromelain	Cysteine	$-p_{-1} \downarrow p_1-$ p_{-1} = non-specific but Lys, Ala, Tyr > others p_1 = non-specific		1–10 pg/µl
Chymotrypsin	Serine	$-p_{-1} \downarrow p_1-$ p_{-1} = aromatic Trp, Tyr, Phe > Leu, Met, Ala p_1 = non-specific.	Active in 0.01% SDS	1–10 pg/µl
Elastase	Serine	$-p_{-1} \downarrow p_1-$ p_{-1} = uncharged, non-aromatic e.g. Ala, Val, Leu, Ile, Gly, Ser, p_1 = non-specific.	Sticks to glass, use plast vials.	1–10 pg/µl

	Type	Specificity	Notes	Concentration
Glu-C	Serine	−Glu ↓ p_1− > −Asp ↓ p_1*	Active in 0.01% SDS	0.05–0.5 ng/μl
Lys-C	Serine	−Lys ↓ p_1−	Active in 0.01% SDS	0.005–0.05 u/μl
Pronase	Mixture	Unspecific		0.5–5 ng/μl
Proteinase K	Serine	$−p_{−1} ↓ p_1$− $p_{−1}$ = non-specific but aromatic or hydrophobic amino acids preferred. p_1 = non-specific.		5–50 pg/μl
Thermolysin	Metallo	$−p_{−1} ↓ p_1 − p_2$ $p_{−1}$ = Leu, Phe > (Ile, Val, Met, Ala) > (Tyr, Gly, Thr, Ser), p_1 = unspecific, p_2 cannot be Pro.	Retains activity at 80°C	5–50 pg/μl
Trypsin	Serine	$−p_{−1} ↓ p_1$− $p_{−1}$ = Lys, Arg p_1 = non-specific	Active in 0.1% SDS.	0.5–5 ng/μl
Subtilisin Carlsbergensis	Serine	$−p_{−1} ↓ p_1$− $p_{−1}$ = non-specific but neutral and acidic amino acids, p_1 = non-specific		0.5–5 ng/μl

*Specificity depends on incubation buffer. In 50 mM ammonium bicarbonate (pH = 7.8) or 50 mM ammonium acetate (pH = 4.0) only Glu bonds are cleaved. In 50 mM phosphate (pH = 7.8) both Asp and Glu bonds are cleaved if p_1 = bulky hydrophobic residue.

Materials

[32]*P-labelled protein* (5 ng/μl in complex buffer)—see Section 3.2.4 for preparation.

RNA ligand[a]

RNA control[a]

RNasin (40 units/μl) (Promega)

E. coli tRNA (RNase free; 5 μg/μl) (Boehringer Mannheim)

BSA (10 μg/μl)

Endoproteinases (Boehringer Mannheim, see Table 4.5)

2 × complex buffer

  20 mM HEPES-KOH, pH 7.9

  200 mM KCl, 4 mM MgCl$_2$

  0.05% Nonidet P-40

RNA renaturation buffer (or equivalent)

  100 mM HEPES-KOH, pH 7.5

  2 mM MgCl$_2$

  0.1 mM EDTA

5 × SDS load buffer

  25% glycerol

  10% SDS

  250 mM Tris-HCl, pH 6.8

  0.025% (w/v) Commassie Blue

  5% 2-mercaptoethanol

30 × 40 cm, 0.5 mm thick, 10–20% Tris-Tricine-SDS-polyacrylamide gel (see Section 4.5.3).

Protein weight marker, e.g. Rainbow™ marker from Amersham.

Equipment

Sequencing gel apparatus

1000 V, 60 mA power supply

37°C and 90°C water bath

Procedure

1. Label a tube for each protein footprinting reaction (or use a microtiter plate).

2. Add 1 μl of *ligand RNA* or *RNA control* in *1 × RNA renaturation buffer*.[a]
3. Add 4 μl of the following mixture to each tube:

   2.5 μl *2 × complex buffer* (or equivalent buffer optimal for complex formation)

   0.7 μl H_2O

   0.5 μl ^{32}P-*labelled protein* (approximately 5 ng/μl in *1 × complex buffer*)[b]

   0.1 μl *RNasin* (40 units/μl)

   0.1 μl *tRNA* (5 μg/μl)

   0.1 μl *BSA* (5 μg/μl)
4. Incubate at 37°C for 15 min.
5. Add 5 μl of *endoproteinase* diluted in *complex buffer*.[c]
4. Incubate at 37°C for 15 min.
5. Place tubes on ice and add 2.5 μl *5 × SDS load buffer*.
6. Heat samples at 90°C for 3 min and load 5 μl on a *30 × 40 cm 10–20% Tris/Tricine-SDS-polyacrylamide gel* (see below). Load also 3 μl *Protein weight marker*.
7. Run samples at a constant current of 40 mA through the stacking gel, followed by 60 mA through the resolving gel (total gel-run is approximately 24 h for 20% gels).
8. Transfer gel to used X-ray film, wrap, and subject to autoradiography with screen at −80°C.

Notes

a. The RNA ligand should be reasonably pure (preferably gel purified) and devoid of proteinases. Best results are generally obtained if the RNA is renatured as described in Section 2.1.2 (Chapter 2). If more samples are tested with the same RNA, the complexes are formed in one tube (steps 2 and 3) before aliquoting. The amount of RNA needed to produce a footprint must be determined empirically. Titrations using 5, 20 and 100 times molar excess of RNA to protein is a good starting point. The same amount of control RNA, known not to bind the protein, should be added to the uncomplexed reaction. Ideally, the control RNA should be produced in the same manner as the ligand RNA. Total tRNA or ribosomal RNA may also be used for many purposes. When mapping mAbs- or protein-protein interactions, the proteins may be diluted in 100 mM Tris-Glycine, pH 7.4 (or appropriate protein dilution buffer) and added at similar molar ratios as described for the RNA.

b. Protein labelling is described in Section 3.2.4. Both N- and C-terminally labelled protein can be probed. ^{33}P-labelled protein may be used if band resolution is critical.

c. To favour single-hit kinetics, at least 50% of the 'mother band' should remain uncleaved. Table 4.5. summarises useful endoproteinases and concentrations which may be used as a starting point for protein footprinting. Generally, the proteases are highly tolerant to buffers and temperature that may be required for stable complex formation, but the activity of proteinases is somewhat batch dependent. Some proteinases also contain RNase activity which will digest the ligand. We have good experience with proteinases from Boehringer Mannheim which also has an extensive collection.

Comments

A major concern in protein footprinting experiments is the risk of secondary cleavages of the protein, which are the result of a primary cleavage elsewhere in the protein. The usage of both N- and C-terminally labelled protein will enable a distinction between primary and secondary cleavages. The criterion for a primary cleavage event is that a particular proteinase cleavage is observed independently of the labelled terminus (see Fig. 4.9 for a similar approach in RNA footprinting methods).

Occasionally, diffuse artefact bands are observed when using trypsin, Arg-C, Lys-C, Glu-C, pronase, proteinase K and in particular thermolysin. The migration of these bands varies dramatically relative to other bands in individual gel runs of the same samples, and these bands are, therefore, easily identified and eliminated for further analysis. The origin of these bands is unknown but they may contain small labelled peptides that migrate with the SDS front in the stacking gel but not in the resolving gel.[20]

Protein fragments are readily identified by comparison to the cleavage pattern obtained with specific proteinases such as Arg-C, Glu-C, Lys-C and Asp-N. Furthermore, the approximately linear correlation between the logarithm of polypeptide mass and gel mobility in a SDS gel allows fairly accurate interpolations between specific size markers. The linear relationship generally is valid for masses above 10 kDa on Tris/Tricine-SDS-polyacrylamide gels. Below 5 kDa, the bands tend to be more closely spaced than expected from their weight differences. Moreover, a nonlinear relationship is occasionally observed for peptides containing long stretches of acidic or basic amino acids, and for unknown reasons abnormally slow migration is observed for the C-terminal cleavage product originating from Asp-N cleavages.

4.5.2. Protein footprinting using Fe^{2+}/H_2O_2

Hydroxyl radicals generated from H_2O_2 in the vicinity of Fe^{2+} can be used to cleave proteins. The advantage of using the Fe^{2+}/H_2O_2 system compared to proteinases is the virtually structure- and sequence-independent cleavages by hydroxyl radicals. Also inaccessibility caused by sterical hindrance of the more bulky proteinases is circumvented. The drawback is that the hydroxyl radicals also react with DNA and RNA and for some RNA-protein complexes we have been unable to obtain a footprint. Moreover, the cleavages result in a smeared pattern probably due to modification of protein side chains. It is therefore crucial that appropriate markers are co-electrophoresed for identification of affected regions. The level of protection is usually weak (less than 2-fold), and the interpretation is greatly facilitated by a PhosphorImager, or similar gel quantification device.

We have used the following procedure to map the binding site of RNA on KH-domain proteins,[23] which is essentially the same protocol described by Heyduk and Heyduk.[13]

Materials
PBS buffer
   20 mM kalium phosphate, pH 7.4
   0.13 M NaCl
ligand RNA (see Sections 4.5.1 Note a and 2.1.2 for preparation
and renaturation, respectively)
[^{32}P]-labelled protein in PBS buffer, see Section 3.2 for preparation
DTT, 50 mM
Fe/EDTA solution
   mixed immediately before use from
   100 μl 10 mM Fe(NH$_4$)$_2$(SO$_4$)$_2$[a]
   100 μl 20 mM EDTA (pH 8.0)
Ascorbate (100 mM)[a]
H$_2$O$_2$ (10 mM)[a]
EDTA (0.2 M), pH 8.0
Thiourea (2 M)
5 × SDS load buffer
   25% glycerol
   10% SDS
   250 mM Tris-HCl, pH 6.8
   0.025% (w/v) Commassie Blue
   5% 2-mercaptoethanol
*30 × 40 cm, 0.5 mm thick, 10–20% Tris-Tricine-SDS-polyacrylam-
ide gel* (see Section 4.5.3)

Equipment
Sequencing gel apparatus
1000 V, 60 mA power supply

Procedure
   1. Label a tube for each protein footprinting reaction (or use a
      microtiter plate).
   2. Add 1 μl of *ligand RNA* or *RNA control* in *PBS buffer*.
   3. Add 4 μl *[^{32}P]-labelled peptide* (approximately 50,000 cpm) in
      PBS buffer and 1 μl 50 mM DTT.

4. Incubate on ice for 15 min.
5. Add:
    2 µl *Fe/EDTA solution*
    1 µl 100 mM *ascorbate*
    1 µl 10 mM H_2O_2 (placed on the lid of the tube)
6. Start reaction by centrifugating the tubes and leave at room temperature for 5 min.
7. During incubation add to the lid of each tube:
    1 µl 2 M *thiourea*
    3 µl 0.2 M *EDTA*
8. Centrifuge briefly to stop reaction.
    Add 3 µl *5 × SDS load buffer*
9. Just prior to loading, heat samples to 70°C for 1 min and load 1–2 µl onto a *30 × 40 cm 10–20% Tris/Tricine-SDS-polyacrylamide gel*. Load also 3 µl *Rainbow Marker* and specific proteinase cleavages as markers.
10. Run samples at constant current of 40 mA through the stacking gel, followed by 60 mA through the resolving gel (total gel-run is approximately 24 h).
11. Transfer gel to used X-ray film, wrap and autoradiograph with screen at −80°C or use PhosphorImager.

Note

a. It is critical that the Fe/EDTA mix, ascorbate and H_2O_2 reagents are prepared immediately before use.

4.5.3. Preparation of Tris/Tricine-SDS-polyacrylamide gels

To enhance the resolution of small protein products we recommend Tris/Tricine-SDS-polyacrylamide gels[24] for the protein footprinting assay.

Materials

Acrylamide mix (30% (w/v) acrylamide, 0.8% (w/v) bisacrylamide). Commercially available as 'Protogel' (Diagnostics).

Gel buffer
   3.0 M Tris-HCl, pH 8.45,
   0.3% SDS
Cathode buffer
   0.1 M Tris base
   0.1 M Tricine
   0.1% SDS
Anode buffer
   0.2 M Tris-HCl, pH 8.9
APS (10%)
TEMED
Isopropanol

Equipment
Sequencing gel apparatus (BRL)
Power supply 1000 V, 60 mA
30 × 40 cm glass plates
0.5 mm spacers and *32-teeth comb*

Procedure
1. Clean *30 × 40 cm glass plates* in water and 96% ethanol. Apply *0.5 mm thick spacers* and seal with tape (or use the rubber sealing system from BRL).
2. Mix resolving gel solution from:
      45 ml *acrylamide mix* (for 20% gel)
      22.5 ml *gel buffer*
      600 μl 10% *APS*
      30 μl *TEMED*
3. Mix and pour the resolving gel slowly, avoiding air bubbles— leave approximately 6 cm space at the top. Overlay with *isopropanol* and leave in a vertical position.
4. When the resolving gel has polymerised, pour off the isopropanol and wash extensively with H_2O.
5. Mix 7% stacking gel solution from:
      3.5 ml *acrylamide mix*

6.5 ml H_2O
5.0 ml *gel buffer*
120 μl 10% *APS*
10 μl *TEMED*

Mix and pour it on top of the resolving gel. Apply *32-teeth comb*.

6. Let the gel polymerise for at least 2 h before use.
7. Place gel in *gel apparatus* and add *cathode buffer* and *anode buffer* to upper and lower buffer chamber, respectively.
8. Carefully remove comb and connect to *power supply*.

4.6. Modification interference

The basis of a modification interference experiment is an initial modification followed by a selection procedure, where RNA molecules that retain their normal properties with respect to the selection procedure are separated from molecules with altered properties (e.g. modified RNA molecules that still bind a protein can be separated from modified RNA molecules that are unable to bind the protein; see Fig. 4.15).

Modification of the RNA prior to the selection procedure lowers the requirements to the modifying chemicals compared with a footprinting experiment (Section 4.3.2). The only requirement is the ability to determine the position of the modification (Fig. 4.16).

The method has been used successfully in studies of protein binding sites on small RNA molecules where the RNA-protein complexes are easily separated from the naked RNA by gel electrophoresis. For larger RNA molecules, where the mobility shift induced by the protein is inadequate for separation by gel electrophoresis, other selection methods such as affinity purification of the complex using antibodies to the protein or affinity tagged proteins can be applied, provided the antibodies or the tag do not interfere with the complex formation (Section 4.3.1).

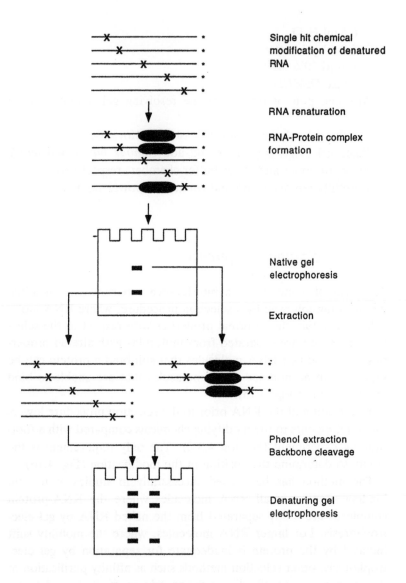

Fig. 4.15. Illustration of the principle of a modification interference experiment.

The method has also been used to identify residues essential for the splicing activity of self-splicing introns where the spliced molecules are easily separated from unspliced RNA.

The RNA can either be modified under physiological conditions (i.e. the RNA molecule is properly folded) or under denaturing conditions (i.e. in the presence of urea or at high temperature). Modifications under physiological conditions will be sensitive to the structure of the RNA molecule, and the modification pattern will not be uniform leaving some residues unmodified. In contrast, modification under denaturing conditions will generate a uniform modification pattern. However, in the latter case the selection will be sensitive to interference both from the folding of the RNA and from the functional behaviour (e.g. binding to a protein).

Hydrazine, CMCT, kethoxal, DMS and DEP (described in Section 4.3.2) are useful for modification interference experiments. Procedures in Section 4.4.2 can be adopted directly for modifications under native conditions, while the procedures used for chemical sequencing of RNA (Section 4.4.3.2.2) generally produce a uniformly modified RNA at denaturing conditions.

The modified residues can be identified by reverse transcriptase analysis as described (Section 4.4.3.1). DMS probing (guanosine at N7), hydrazine and DEP modification can also with advantage be applied to end-labelled RNA, where the selected RNA is cleaved by acidic aniline strand scission to identify the modified residues as described in Section 4.4.3.2.2.

4.6.1. Base modifications

Hydrazine   Probing either **U or C > U** (Section 4.4.3.2.2). Modified residues can be detected by reverse transcriptase methods or end-labelled methods after an acidic aniline cleavage (Section 4.4.3). Hydrazine modifications can only be performed under denaturing conditions.

DEP   **A > G (N7)** under denaturing conditions (Section

	4.4.3.2.2) or **A (N7)** under native conditions (Section 4.4.2.3). Modified residues can be detected directly by reverse transcriptase methods or after acidic aniline cleavage of end-labelled species.
DMS	Probing **G (N7)**, **A (N1)** and **C (N3)**. (Section 4.4.3.2.2, denaturing conditions, or Section 4.4.2.2, native conditions). Modification at A (N1) and C (N3) can be identified by reverse transcriptase methods while identification of G (N7) modifications by reverse transcriptase methods or end-labelled methods (Section 4.4.3) requires a borohydride-reduction followed by an acidic aniline cleavage (Section 4.4.2.2).
Kethoxal	Probing **G (N1 and N2)**. (Section 4.4.2.4, native conditions, or follow Section 4.4.2.4, denaturing conditions), but incubate for 2 min at 95°C. Modified residues can be detected directly by the reverse transcriptase method.
CMCT	Probing **U (N3) > G (N1)**. (Section 4.4.2.5, native conditions or follow Section 4.4.2.4, denaturing conditions), but incubate for 5 min at 95°C. Modified residues can be detected directly by reverse transcriptase methods.

4.6.2. Phosphate modification by ethylnitrosourea (ENU)

Ethylnitrosourea (ENU; Sigma) ethylates the phosphodiester bond of RNA forming relatively unstable phosphotriester bonds. Subsequent incubation at increased pH results in a preferential cleavage at ethylated phosphates. The RNA is ethylated under denaturing condition,[26,27] refolded, and selected for function or binding ability.

Materials
2 × modification buffer
  600 mM Na cacodylate, pH 8.0
  10 mM EDTA

Strand scission buffer
    100 mM Triethylammoniumbicarbonate, pH 9.0
    (Adjust pH by bubbling CO_2 through the solution)
ENU solution[a] (saturated solution in ethanol) *carrier tRNA*
(5 μg/μl)

Procedure[b]
Modification
 1. Mix
        9 μl H_2O containing 50 ng (approximately 10^6 cpm) 5'- or
        3'-end labelled RNA.
        1 μl *carrier tRNA*.
        10 μl *2 × modification buffer*
 2. Add
        5 μl of a *ENU solution*
 3. Incubate 2 min at 80°C.
 4. Add 175 μl of 0.25 M NaOAc pH 6.0 and precipitate with 500
    μl EtOH (−20°C).
 5. Wash pellet with 100% ethanol.
 6. Resuspend RNA in appropriate buffer.
Selection[c]
 7. Renature the RNA by heating to 37°C for 5 min.
 8. Add protein at a concentration which results in <50% com-
    plexes.
 9. Separate bound RNA from unbound RNA in a native poly-
    acrylamide gel as described in Section 4.2.3.
10. Excise the bands and elute the RNA as described in Section
    2.1.1 of Chapter 2.
11. Extract once with phenol and once with chloroform.
12. Add 10 μg of *carrier tRNA* to each sample.
13. Precipitate with 1/10 vol of 0.3 M NaOAc and 2.5 vol of EtOH.
14. Wash with 80% EtOH (−20°C) and dry.
Strand scission
15. Resuspend in 10 μl *strand scission buffer*.

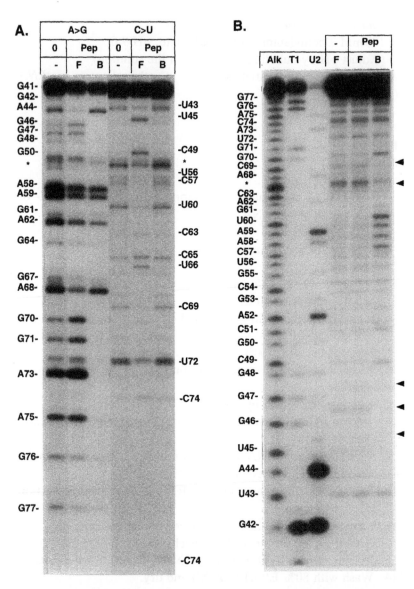

Fig. 4.16. Autoradiograms showing a chemical interference analysis of RNA using base modifications (A), or phosphate modifications (B). A 37-nucleotide long RNA

16. Heat at 50°C for 5 min to cleave RNA at ethylated phospho-triester bonds.
17. Lyophilise.
18. Resuspend in urea load buffer.[d]

Notes

a. Store solid ENU (Sigma) at 4°C in the dark. Make a saturated solution in ethanol (approximately 750 mM) immediately before use. ENU is highly toxic and must be handled with extreme care. Treat ENU waste with 10 M NaOH before disposal.

b. The procedure describes ENU modification performed on end-labelled RNA selected for its binding affinity to a protein, but a similar procedure can be used for other types of selection and with 'cold' RNA followed by reverse transcriptase analysis.

c. Ethylated RNA is relatively unstable even at neutral pH. Therefore, it is important to perform the native gel electrophoresis immediately after modification and for as short a time as possible. Denaturing gel purification of ethylated RNAs prior to the native gel run results in massive degradation and is not recommended.

d. It is important to prepare and co-electrophorese a hydrolysis ladder and some sequence-specific reactions (Section 4.4.3). Note that cleavages induced by ENU ethylation leave the phosphate on the 5'-end of the RNA product whereas enzymatic and

---

encompasing the high affinity binding site for HIV–1 Rev protein was either 3' end-labelled (A), or 5' end-labelled (B), and treated with either DEP (for A > G, see Section 4.4.3.2.2), or hydrazine/NaCl for (C > U, 4.4.3.2.2) (A), or ENU (Section 4.6.2). Free RNA (F) and RNA bound to Rev protein (B) were separated and purified by native gel electrophoresis. Nonselected but modified RNA is denoted '–'. The isolated RNA is cleaved by acidic/aniline treatment and analysed by 15% denaturing PAGE. Alk, T1 and U2 indicate alkaline hydrolysis, G-specific RNase T1 digest and A-specific RNase U2 digest, respectively. Arrows indicate positions at which phosphate ethylation interferes with protein binding. Note the reduction in charge inferred by the ethylation causes a slower migration of the ENU modified RNA compared to the sequence ladder. The star indicates a strong stacking of nucleotides hindering a detailed assignment.

alkaline hydrolysis produce a 3'-end phosphate. Consequently, the 5'-end labelled RNA products from the ENU treatment migrate slightly slower than the marker sequence and the 3'-end labelled products migrate faster.

4.6.3. Analogue interference

An alternative to post-transcriptional modifications of the RNA is co-transcriptional modification by addition of modified nucleotides during the *in vitro* transcription. T7 RNA polymerase incorporates a number of modified residues, including modifications of the phosphate group, the ribose or/and the base moiety (Section 2.3.2.2). In an analogue interference experiment it is mandatory that the RNA can be cleaved at the modified site subsequent to the selection step. Phosphorothioate substitution, where a nonbridging Rp oxygen is replaced by a sulphur, provides an easy method for selective cleavage by iodine.[28,29] Concurrent incorporation of a phosphorothioate and an additional modification of the same residue provide a quantitative evaluation of every chemical group in the RNA molecule provided it can be incorporated by the polymerase. Combinations of phosphorothioate and ribose modifications as 2'-O-methoxy- and deoxy derivatives[30] and combinations with base substitutions as biotin-8-adenosine, methyl-7-guanosine, inosine or diaminopurine[31,32] have been reported.

Procedure

1. An appropriate selection procedure is applied to the RNA transcript containing the modified residues.
2. The selected pool of RNA is then cleaved quantitatively at the modified residues by iodine as described in Section 4.4.2.7 and analysed by polyacrylamid gel electrophoresis.

4.7. SELEX

Systematic evolution of ligands by exponential enrichment (SELEX), or *in vitro* selection, is a method developed in the laboratories of Gerald Joyce, Larry Gold, Jack Szostak and coworkers in the late 1980s.[33-35] It is a powerful method to screen a large population (approximately 10^{15}) of independent oligonucleotides for a particular function, e.g. binding of proteins and small ligands or generating ribozymes (reviewed in Refs. 36–39). The method involves three major steps: (a) Generation of a large pool of diverse oligonucleotides; (b) Selection of a subset of molecules with a particular property using various selection techniques; and (c) Amplification of the selected molecules. At this stage the selected molecules may either be characterised by sequencing or used as input in an additional selection cycle (see Fig. 4.17). The cycle is repeated a variable number of times depending on the system.

The ease of forming and purifying RNA-protein complexes *in vitro* has made SELEX a generally useful method to study protein-RNA interactions. Selection of an RNA substrate for a particular protein is readily performed using standard techniques, such as filter binding (Section 4.2.1), mobility shift analysis (Section 4.2.3), or affinity chromatography (Section 4.3.1). Even if the protein does not bind RNA as part of its natural *in vivo* function it is feasible to select nucleic acid ligands which may be useful for therapeutics or industrial purposes. In studies of proteins that naturally interact with RNA targets, the SELEX approach is useful in the analysis of RNA sequence specificity of the protein.

The randomised nucleotide library that is used as the input in the initial selection procedure is conveniently generated on a DNA synthesiser and transcribed into an RNA pool by *in vitro* transcription. Designing the structure of the library is a crucial step which should be planned carefully. Three main types of libraries are in general use: A fully randomised library; a randomised region which is embedded in a natural RNA structure; and a 'doped' library where each position of a specific RNA sequence contains a

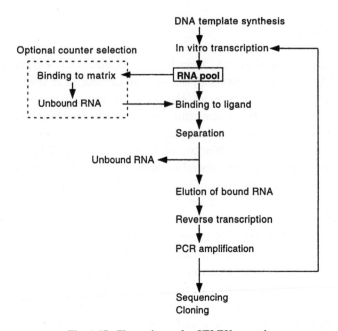

Fig. 4.17. Flow scheme for SELEX procedure.

variable amount of other nucleotides. The fully randomised library usually yields sequences with no resemblance to the natural target, either on the primary or secondary level. This may reflect that these RNAs either interact with different sites of the protein or in a structural unrelated fashion. In the embedded and 'doped' libraries the selected RNAs often show some resemblance with the natural substrate, which makes this approach more favourable in studies of specificity determinants in naturally occurring RNA-protein interactions.

An additional consideration is the required level of library complexity and RNA length. Due to experimental limitations the pool of RNAs has a practical limit of approximately 10^{15} molecules. This corresponds to approximately the number of sequence combinations in a randomised 25-nucleotide long oligonucleotide ($4^{25} \simeq 10^{15}$). However, for some applications it is desirable to randomise considerably larger oligonucleotides. A library containing 220 nucleotides was successfully used to study catalytic RNAs.[40] In such a study the number of theoretical combinations is 4^{220} which for a fully represented library would exceed the mass of the universe. The potential homology among the selected molecules is therefore limited to only a fraction of the nucleotides and more elaborate sequence analysis is required. A general scheme for construction of the library and amplification of the selected RNAs is essentially based on the Tuerk and Gold procedure[35] and outlined in Fig. 4.18.

Materials
DNA oligonucleotides, preferably HPLC or gel purified:
    DNA oligonucleotide 1, 2 and 3 (see Fig. 4.18).
8% denaturing polyacrylamide gel
Nitrocellulose filter: (BA85, Schleicher & Schuell)
2 × Renaturation buffer
    20 mM HEPES-KOH, pH 7.0
    2 mM $MgCl_2$
    200 mM KCl
Washing buffer[a]
Filter elution buffer
    300 mM NaCl
    1% SDS
    10 mM Tris-HCl, pH 7.5
    1 mM EDTA
RNA precipitation buffer
    0.3 M NaOAc, pH 6.0
    1 mM EDTA
10 × Annealing buffer
    100 mM Tris-HCl, pH 8.0
    500 mM KCl
10 × Reverse transcription buffer
    400 mM Tris-HCl, pH 8.0
    60 mM $MgCl_2$

pT7 →
5'-TAATA CGACT CACTA TAGGG AGGCA ACGCC ACAAT TCCGA TCAAG-3'   oligo 1
3'-ATTAT GCTGA GTGAT ATCCC TCCGT TGCGG TGTTA AGGTT AGTTC NNNNN NNNNN NNNNN NNNNN NNTAG ATACT TTCTT AAAT ATAGA GATAA CTTTG-5'   oligo 2

⇓ transcription

5'-GGG AGGCA ACGCC ACAAU UCCGA UCAAG NNNNN NNNNN NNNNN NNNVC UAUGA AAGAA UUUA UAUCU CUAUU GAAC-3'
                                                                         G AUACT TTCTT AAAT ATAGA GATAA CTTTG-5'   oligo 3

⇓ selection, reverse transcription

5'-TAATA CGACT CACTA TAGGG AGGCA ACGCC ACAAT TCCGA TCAAG-3'   oligo 1
3'-CCC TCCGT TGCGG TGTTA AGGTT AGTTC NNNNN NNNNN NNNNN NNNNN NNTAG ATACT TTCTT AAAT ATAGA GATAA CTTTG-5'   oligo 3

⇓ amplification for subsequent rounds or cloning

Fig. 4.18. Principles for construction and amplification of a library. The *DNA oligonucleotide 2* contains: The template strand of the T7 RNA polymerase promoter, a fixed 28-nucleotide region (5'-tag), the randomised region of desired length, and a fixed 28-nucleotide region (3'-tag). If the randomised insert is larger than 30 nucleotides it is recommended to synthesise oligonucleotide 2 in three parts and perform a double ligation using bridging oligonucleotides. *DNA oligonucleotide 1* corresponds to the complementary strand of the T7 RNA polymerase promoter and the 5'-tag (transcription by T7 RNA polymerase requires that the promoter is double stranded and that the first 1–2 nucleotides of the RNA strand are guanines). After transcription and selection, the RNA products are reverse transcribed into cDNA using *DNA oligonucleotide 3*, containing complimentary sequence to the 3'-tag. The cDNA is amplified using oligonucleotides 1 and 3 yielding a template for 2. round of RNA transcription or for cloning into bacterial plasmid.

50 mM DTT
AMV reverse transcriptase (20 units/μl)
dNTP mix (5 mM of each dATP, dGTP, dCTP, dTTP)
10 × PCR buffer
   100 mM Tris-HCl, pH 9.0
   500 mM KCl
   15 mM MgCl$_2$
Taq DNA polymerase (5 units/μl)
Low-melting agarose (NuSieve, FMC)

Equipment

G-50 Sepharose flow column with UV detector
PCR cycler
Vacuum filter blotter

Procedure

Constructing the library (see Fig. 4.18).

1. Mix 1 nmol of both *oligonucleotides 1 and 2* in 200 μl of TE buffer.
2. Anneal oligonucleotides by incubation at 50°C for 5 min.
3. Use annealed oligonucleotides as template in a 1-ml transcription reaction according to Section 2.3.1 of Chapter 2.[b]
4. Purify the RNA on a preparative *8% denaturing polyacrylamide gel* as described in Section 2.1.2 of Chapter 2.
5. Resuspend purified RNA at 20 pmol/μl H$_2$O.
6. Measure UV absorption at 260 nm to determine the concentration (see Section 2.1.2 of Chapter 2). Yield should be in the range of 1–10 nmol of RNA.

Selecting RNA molecules (1st round)

1. Mix 50 μl (1 nmol) of the RNA library with 50 μl *2 × renaturation buffer*.
2. Denature the RNA by incubating at 70°C for 3 min, and renature at 4°C for 5 min before selection.
3. Adjust the buffer to appropriate binding conditions, if necessary.[c]

4. Mix with 10 pmol of protein in 100 μl of the same buffer.
5. Incubate at 20–37°C for 1–10 min.
6. Pre-wet filter in *washing buffer.*[a]
7. Let the mixture pass through a *nitrocellulose filter* using a *vacuum filter blotter* and rinse with 10 ml *washing buffer*.
8. Elute the RNA by incubating the filter in 400 μl *filter elution buffer*.
9. Extract the recovered RNA once with phenol/chloroform, once with chloroform and precipitate.
10. Redisolve in 50 μl H_2O.

Amplification of selected RNAs

1. Perform reverse transcription by mixing:
   40 μl eluted RNA
   1 μl of *DNA oligonucleotide 3* (100 pmol/μl)
   5 μl *10 × annealing buffer*
2. Incubate at 65°C for 5 min and transfer to ice.
3. Add:
   5 μl 5 mM *dNTP mix*
   6 μl *10 × reverse transcription buffer*
   1 μl *AMV reverse transcriptase* (10 units)
4. Incubate at 42°C for 15 min.
5. PCR amplification of first strand synthesis is performed in a 500 μl volume containing:
   340 μl H_2O
   50 μl cDNA (from step 4)
   50 μl *10 × PCR buffer*
   10 μl of DNA oligonucleotides 1 and 3 (100 pmol/μl)
   50 μl 5 mM *dNTP mix*
   3 μl Taq DNA polymerase
6. Divide into 10 reactions (50 μl each) and perform 25 PCR cycles as follows: 94°C 1 min, 60°C 1 min, 72°C 1 min in a *PCR cycler*. The number of cycles may be reduced gradually to 15 in the final round.
7. The PCR product is purified by low-melting agarose electrophor-

esis and either used as template in a new round of RNA transcription (8a) or cloned into a plasmid for sequencing (8b).

8a. Transcribe RNA from the PCR fragment in a 200 μl reaction
according to Section 2.3.1 (Chapter 2).

8b. To clone the PCR products it is convenient to use the TA
cloning procedure which takes advantage of the extruding A
residue of Taq amplified PCR fragments.[d] The ligated plasmids
are transformed into bacteria, and plasmids from individual
colonies are sequenced.

Selection of RNA in subsequent rounds[e]
Use only 100–200 pmol RNA and 10 pmol of protein in 100 μl
reactions

Notes

a. Washing buffers should be optimised. Many procedures use very
simple washing buffers (e.g. 50 mM Tris-OAc).

b. Ideally, this volume will support the synthesis of about 10^{15}
RNA molecules. The size of the transcription mixture may be
reduced if the complexity of the library is not critical, or if the
randomised sequence is shorter. Test the transcription efficiency
initially in a 100 μl reaction. Moreover, it is important to treat
the RNA with DNAse (RNase free) to remove the template.

c. The binding conditions in the selection step is extremely important. Preferably conditions close to physiological conditions
should be chosen. Tris-HCl or potassium phosphate buffers (pH
7.0–7.5), Mg^{2+} ions, unspecific RNA and nonionic detergents
(Triton-X 100 and Tween-20) are commonly used. The binding
conditions range from a few minutes to several hours at tempera-
tures between 0–37°C.

d. Alternatively one may include restriction cleavage sites in the
constant region of the primers which may be used for cloning.
However, cleaving the pool of amplified DNA fragments with
restriction enzymes will eliminate sequences in the library containing their recognition sequences.

e The number of rounds required is variable and must be determined from sequencing and functional analysis of the pools obtained. Increasing the number of rounds reduces the diversity of the final nucleic acid pool, thereby increasing the likelihood of picking the best candidates. However, some diversity in the final pool is desirable in order to create alignments for structure prediction (see Ref. 41, for an excellent example of phylogenetic secondary structure prediction). A high background in the selection step tends to increase the number of required rounds. However, this may also increase the risk of selecting artefact RNA molecules with affinity for the employed matrix such as a nitrocellulose filter.

Comments

It is possible to expand the number of RNA molecules in an ongoing selection process by altering the conditions during the PCR step, causing an increase in the mutation rate.[42] Mutagenic PCR is used to introduce mutations into already selected sequences and is therefore a powerful technique for sampling the local sequence space.

To reduce the risk of selecting artefact RNA molecules with affinity to the matrix many procedures include a counter-SELEX against the partitioning matrix. The oligonucleotides are simply passed through a selection step using the same matrix but without the ligand. Only the unbound pool of RNAs (the flow-through) is used in the subsequent selection step (Fig. 4.17).

References

1. Spierer, P., Bogdanov, A.A. and Zimmermann, R.A. (1978). Parameters for the interaction of ribosomal proteins L5, L18 and L25 with 5S RNA from *Escherchia coli*. Biochemistry *17*, 5394–5398.
2. Griffith, O.M. (1986). Techniques of Preparative, Zonal and Continuous Flow Ultracentrifugation. Beckman Instruments, Palo Alto, California.

3. Rickwood, D. (1992). Preparative centrifugation—A practical approach. In: The Practical Approach Series, Vol. 113 (Rickwood, D. and Hames, B.D, eds.). Oxford University Press, Oxford.

4. Lamond, A.I. and Sproat, B.S. (1994). Isolation and characterization of ribonucleoprotein complexes. in RNA Processing, Vol. I (Higgens, S.J. and Hames, B.D, eds.). IRL Press, Oxford, pp. 103–140.

5. Dontsova, O.A., Rosen, K.V., Bogdanova, S.L., Skripkin, E.A., Kopylov, A.M. and Bogdanov, A.A. (1992). Identification of the *Escherichia coli* 30S ribosomal subunit protein neighboring mRNA during initiation of translation. Biochimie *74*, 363–371.

6. Sergiev, P.V., Lavrik, I.N., Wlasoff, V.A., Dokudovskaya, S.S., Dontsova, O.A., Bogdanov, A.A. and Brimacombe, R. (1997). The path of mRNA through the bacterial ribosome: a site-directed cross-linking study using new photoreactive derivatives of guanosine and uridine. RNA *3*, 464–475.

7. Rana, T.M. and Meares, C.F. (1991). Transfer of oxygen from an artificial protease to peptide carbon during proteolysis. Proc. Natl. Acad. Sci. USA *88*, 10578–10582.

8. Heilek, G.M. and Noller, H.F. (1996). Site-directed hydroxyl radical probing of the rRNA neighborhood of ribosomal protein S5. Science *272*, 1659–1662.

9. Gish, G. and Eckstein, F. (1988). DNA and RNA sequence determination based on phosphorothioate chemistry. Science *240*, 1520–1522.

10. Donis-Keller, H. (1979). Site specific enzymatic cleavage of RNA. Nucleic Acids Res. *7*, 179–191.

11. Peattie, D.A. (1979). Direct chemical method for sequencing RNA. Proc. Natl. Acad. Sci. USA *76*, 1760–1764.

12. Zueva, V.S., Mankin, A.S., Bogdanov, A.A. and Baratova, L.A. (1985). Specific fragmentation of tRNA and rRNA at a 7-methylguanine residue in the presence of methylated carrier RNA. Eur. J. Biochem. *146*, 679–687.

13. Heyduk, E. and Heyduk, T. (1994). Mapping protein domains involved in macromolecular interactions: A novel protein footprinting approach. Biochemistry *33*, 9643–9650.

14. Ellgaard, L., Holtet, T.L., Moestrup, S.K., Etzerodt, M. and Thogersen, H.C. (1995). Nested sets of protein fragments and their use in epitope mapping: characterization of the epitope for the S4D5 monoclonal antibody binding to receptor associated protein. J. Immunol. Meth. *180*, 53–61.

15. Cohen, S.L., Ferré-DAmaré, A.R., Burley, S.K. and Chait, B.T. (1995). Probing the solution structure of the DNA-binding protein Max by a combination of proteolysis and mass spectrometry. Protein Sci. *4*, 1088–1099.

16. Jensen, T.H., Leffers, H. and Kjems, J. (1995). Intermolecular binding sites of HIV-1 Rev protein determined by protein footprinting. J. Biol. Chem. *270*, 13777–13784.

17. Hori, R., Pyo, S. and Carey, M. (1995). Protease footprinting reveals a surface on transcription factor TFIIB that serves as an interface for activators and coactivators. Proc. Natl. Acad. Sci. USA *92*, 6047–6051.

18. Zhong, M., Lin, L. and Kallenbach, N.R. (1995). A method for probing the topography and interactions of proteins: Footprinting of myoglobin. Proc. Natl. Acad. Sci. USA *92*, 2111–2115.

19. Jensen, T.H., Jensen, J., Szilvay, A.M. and Kjems, J. (1997). Probing the structure of HIV-1 Rev by protein footprinting of multiple monoclonal antibody binding sites. FEBS Lett. *414*, 50–54.

20. Lykke-Andersen, J., Garrett, R.A. and Kjems, J. (1996). Protein footprinting approach to mapping DNA binding sites of two archael homing enzymes: evidence for a two-domain protein structure. Nucleic Acids Res. *24*, 3982–3989.

21. Tange, T.O., Jensen, T.H. and Kjems, J. (1996). In vitro interaction between human immunodeficiency virus type 1 Rev protein and splicing factor ASF/SF2-associated protein, p32. J. Biol. Chem. *271*, 10066–10072.

22. Heyduk, T., Heyduk, E., Severinov, K., Tang, H. and Ebrigth, R.H. (1996). Determinants of RNA polymerase α subunit for interaction with β, β' and σ subunits: Hydroxyl-radical protein footprinting. Proc. Natl. Acad. Sci. USA *93*, 10162–10166.

23. Leffers, H., Jensen, T.H., Andersen, A. and Kjems, J. (1998). Mapping the RNA binding sites within the KH-domain of PCBP-1 and -2 (submitted).

24. Schagger, H. and von Jagow, G. (1987). Tricine-sodium dodecyl sulfate-polyacrylamide gel electrophoresis for the separation of proteins in the range from 1 to 100 kDa. Anal. Biochem. *166*, 368–379.

25. Jensen, T.H., Jensen, A. and Kjems, J. (1995). Tools for expression and purification of full length, N- or C-treminal ^{32}P labeled protein, applied on HIV-1 Gag and Rev. Gene *162*, 235–237.

26. Vlassov, V.V., Giege, R. and Ebel, J.P. (1981). Tertiary structure of tRNAs in solutions monitored by phosphodiester modification with ethylnitrosourea. Eur. J. Biochem. *119*, 51–59.

27. Kjems, J., Calnan, B.J., Frankel, A.D. and Sharp, P.A. (1992). Specific binding of a basic peptide from HIV-1 Rev. EMBO J. *11*, 1119–1129.

28. Waring, R.B. (1989). Identification of phosphate groups important to self-splicing of the Tetrahymina rRNA intron as determined by phosphorothioate substitution. Nucleic Acids Res. *17*, 10281–10293.

29. Gish, G. and Eckstein, F. (1988). DNA and RNA sequence determination based on phosphorothioate chemistry. Science *240*, 1520–1522.

30. Conrad, F., Hanne, A., Gaur, R.K. and Krupp, G. (1995). Enzymatic synthesis of 2'-modified nucleic acids: identification of important phosphate and ribose moieties in RNase P substrates. Nucleic Acids Res. *23*, 1845–1853.

31. Kahle, D., Wehmayer, U., Char, S. and Krupp, G. (1990). The methylation of one specific guanosine in a pre-tRNA prevents cleavage by RNase P and by the catalytic M1 RNA. Nucleic Acids Res. *18*, 837–844.

32. Strobel, S.A. and Shetty, K. (1997). Defining the chemical groups essential for Tetrahymena group I intron function by nucleotide analog interference mapping. Proc. Natl. Acad. Sci. USA *94*, 2903–2908.

33. Joyce, G.F. (1989). Amplification, mutation and selection of catalytic RNA. Gene *82*, 83–87.

34. Ellington, A.D. and Szostak, J.W. (1990). In vitro selection of RNA molecules that bind specific ligands. Nature *346*, 818–822.

35. Tuerk, C. and Gold, L. (1990). Systematic evolution of ligands by exponential enrichment: RNA ligands to bacteriophage T4 DNA polymerase. Science *249*, 505–510.

36. Gold, L., Polisky, B., Uhlenbeck, O. and Yarus, M. (1995). Diversity of oligonucleotide functions. Ann. Rev. Biochem. *64*, 763–797.

37. Szostak, J.W. and Ellington, A.D. (1993). In vitro selection of functional RNA sequences. In: RNA World (Gesteland, R.F. and Atkins, J.F., eds.). Cold Spring Harbor Laboratory Press, New York, pp. 511–534.

38. Conrad, R.C., Giver, L., Tian, Y. and Ellington, A.D. (1996). In vitro selection of nucleic acid aptamers that bind proteins. Meth. Enzymol. *267*, 336–367.

39. Eaton, B.E. and Pieken, W.A. (1995). Ribonucleosides and RNA. Ann. Rev. Biochem. *64*, 837–863.

40. Bartel, D.P. and Szostak, J.W. (1993). Isolation of new ribozymes from a large pool of random sequences. Science *261*, 1411–1418.

41. Ekland, E.H., Szostak, J.W. and Bartel, D.P. (1995). Structurally complex and highly active RNA ligases derived from random RNA sequences. Science *269*, 364–370.

42. Beaudry, A.A. and Joyce, G.F. (1992). Directed evolution of an RNA enzyme. Science *257*, 635–641.

CHAPTER 5

Functional analysis of RNA-protein complexes *in vitro*

5.1. Pre-mRNA splicing

Pre-mRNA splicing is accomplished by the spliceosome that is assembled from five small nuclear ribonucleoproteins particles (U1, U2, U4, U5 and U6 snRNPs) and a large number of protein splicing factors (see Moore and Sharp for review).[1] The assembly process is directed by specific intermolecular interactions between these factors and with the RNA sequences surrounding the exon-intron boundaries, the 5'- and 3'-splice sites. *In vitro* splicing of exogenous RNA transcripts containing introns was first described by Grabowski et al.[2] and Hernandez and Keller[3] using nuclear extracts prepared according to Dignam et al.,[4] and these procedures are the standard approach today. The nuclear extract can be further fractionated by chromatography into six fractions that combined yield splicing activity.[5] Moreover, the cytoplasmic S100 fraction supplemented with purified SR-proteins can be used as a source of competent splicing factors (see Section 3.1.2).

For efficient splicing constructs the *in vitro* splicing is in the 50–80% range. However, some transcripts that are spliced inefficiently *in vivo*, are often generating a significant amount of aberrant splicing products *in vitro*, probably due to prevalent use of pseudo-splice sites. Also the size of the intron has a large effect on the splicing efficiency *in vitro*. Whereas introns of up to 200,000 nucleotides are spliced efficiently *in vivo* the practical limit *in vitro* is about 2000–3000 nucleotides. Primary transcripts containing more than one intron can also be assayed *in vitro*, but interpretation of the data rapidly becomes complex. There are many examples in the literature demonstrating that *cis*-acting regulatory elements, such as splicing enhancers or silencers can be studied in *in vitro* assays, suggesting that the factors necessary for this cross-talk generally are present in a nuclear extract in sufficient amounts.

Traditionally, a splicing construct contains an intron flanked by two exons. However, interactions across the exons play an important role in spliceosome assembly, a phenomenon termed exon definition.[6] The presence of an upstream 3'-splice site generally stimulates the function of a weak downstream 5'-splice site. Similarly, a downstream 5'-splice site induces the usage of a weak upstream 3'-splice site. In this transactivation model, the 5'-cap structure of the first exon and the polyadenylation site in the terminal exons act as 3'- and 5'-splice sites, respectively, in promoting the spliceosome assembly. This view is reinforced by the observation that less efficiently spliced introns may be stimulated by moving the 5'-cap closer to the 5'-splice site or by inserting a functional 5'-splice- or polyadenylation site in the 3'-exon. However, exon definition seems only to play a minor role for efficiently spliced introns *in vitro*.

Although the splicing procedure below works well for most introns, the efficiency can generally be improved considerably by titrating the ionic conditions as discussed below.

5.1.1. In vitro *mRNA splicing*

Materials
Uniformly [32]*P-labelled GpppG capped pre-mRNA*[a]
ATP (10 mM)
Creatine phosphate (50 mM)
MgCl₂ (50 mM),
Bulk E. coli tRNA: 5 µg/µl (RNase free from Boehringer Mannheim)
Nuclear extract: from HeLa cells prepared as described in Section 3.1.1.[b]

Buffer D
  20 mM HEPES-KOH, pH 7.5
  100 mM KCl
  0.2 mM EDTA
  1 mM DTT
  10% glycerol
Splice stop buffer
  0.5% SDS
  5 mM EDTA
  0.1 M NaCl
  20 mM Tris-HCl, pH 8.0
  0.3 M NaOAc (pH 6.0)
  25 μg/ml *E. coli* tRNA
RNA load buffer
  80% formamide, v/v
  0.1% xylene cyanol
  0.1% bromophenol blue
  1 mM EDTA, pH 8.0

Procedure

1. Clear the *nuclear extract* after thawing by centrifugation at 10,000 rpm for 1 min. Transfer the supernatant to a new tube; keep on ice.[b]

2. For a standard splicing reactions (20 μl) mix the following ingredients on ice:[c]

     2 μl 10 mM ATP
     2 μl 50 mM *creatine phosphate*
     1 μl 50 mM *MgCl$_2$*[d]
     1 μl 5 μg/μl bulk *E. coli* *tRNA*[e] (optional)
     7 μl *nuclear extract* (containing approximately 10 μg/μl protein in *buffer D*)[f]
     3 μl *buffer D*
     $1-5 \times 10^4$ cpm of *uniformly ^{32}P-labelled GpppG capped pre-mRNA* and adjust with H_2O to 20 μl.

3. Mix thoroughly by flicking the tube and spin down.

4. Incubate for 20–180 min at 30°C.[g]
5. Transfer 5 µl of splicing reaction to new tube if splicing complexes are to be analysed on native gels (Section 5.1.3).
6. Stop the splicing reaction by adding 180 µl of *splice stop buffer*, vortex vigorously for 15 s.
7. Extract once with 1× phenol/chloroform, transfer 140–160 µl of the water phase to a new tube. Make sure not to take any of the thick protein interphase.
8. Extract water phase once more with chloroform and transfer water phase to a new tube.
9. Precipitate RNA with 2.5 vol. EtOH. Generally 20–50% of the radioactivity will remain in the EtOH after precipitation, which probably are small RNAs and nucleotides originating from RNA degradation.
10. Redissolve pellet in *RNA load buffer*. Store at −80°C if not loaded on gel immediately.

Notes

a. Preparation of the RNA template: RNA may conveniently be synthesised by *in vitro* transcription (Section 2.3.2). Capping the transcripts with GpppG greatly stabilises the RNA and enhances the splicing. However, it is unnecessary to use the more expensive di-methylated cap functional *in vivo*. Presumably, the cap is rapidly methylated in nuclear extract. The mRNA can be stabilised by inserting a 10–12 bp stable hairpin structure right at the 3′-end. If the RNA is gel purified before splicing it is important not to have any SDS present during the extraction from the gel slice since the splicing reaction is highly sensitive to trace amounts of SDS. Store RNA in water at −80°C until use.

b. HeLa cell nuclear extracts are prepared as described by Dignam et al.[4] (Section 3.1.1), and stored in aliquots at −80°C. At this temperature it can be stored almost indefinitely without loosing splicing activity. The splicing activity of some preparations of nuclear extracts may be improved by pre-incubating the extract for 10 min at 30°C in the presence of 1 mM ATP. Unused extract can be quick-frozen and stored at −80°C with no apparent decline in splicing activity.

c. A negative control for the splicing reaction is to remove ATP: Preincubate the nuclear extract for 20 min at 30°C to deplete the endogenous pool of ATP. Mix the splicing reaction in the absence of ATP and creatine phosphate. Otherwise incubate and treat the control as other samples.

d. Optimal $MgCl_2$ concentration may vary between RNA substrates. 2.5 mM are optimal for most purposes (PIP and Adenovirus constructs, use 1.7 mM).

e. Extracts contain an exonuclease activity which removes 3′-end nucleotides with variable efficiency. This activity is inhibited by adding 0.25–0.5 mg/ml bulk *E. coli* tRNA (RNase free) to the splicing reaction. More than 1 mg/ml of tRNA will inhibit the splicing reaction. In many procedures, 1 unit/μl RNasin (final concentration) is added to the splicing reaction. We find that it generally makes no difference.

f. The amount of nuclear extract added can be varied (4–12 μl). Make sure that the KCl concentration always is between 30–90 mM by adjusting with buffer D (100 mM KCl). The optimal concentration may vary and should be adjusted for each batch of extract and for different mRNAs. A total of 50 mM KCl (nuclear extract + buffer D) seem optimal for most purposes. Less efficiently spliced introns may need additional SR proteins to detect splicing products *in vitro*.

g. Splicing products from efficiently spliced mRNAs generally appear after 15–20 min. Splicing yield usually increases up to 2–3 h of incubation. Longer incubation generally does not increase the yield of splicing products mainly due to the unspecific degradation of RNA.

5.1.2. Analysis of RNA splicing products by denaturing gel electrophoresis

The analysis and quantification of splicing products are conveniently done by denaturing gel electrophoresis. An example of the interpretation of a 'splicing gel' is shown in Fig. 5.1.

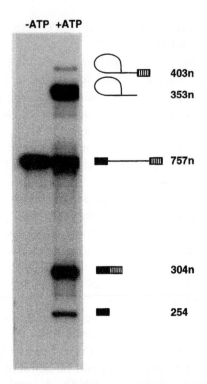

Fig. 5.1. Autoradiogram of a denaturing polyacrylamide splicing gel. Splicing products from a 757 nucleotide transcript that is a modified version of human β-globin pre-mRNA containing a 353 nucleotide intron. The splicing products are analysed in a 6% polyacrylamide, 0.75 × TBE gel and their identity and length are indicated. Filled boxes, cross-hatched boxes and thin lines denote the 5'- and 3'-exons and intron, respectively. Note that the splicing is strictly dependent on addition of ATP as co-factor and that the lariat bands migrate considerably slower than expected from their size.

Material

Denaturing gel

    4–15% polyacrylamide depending on RNA products[a]

8 M urea

    0.5–1.0 × *TBE buffer* depending on RNA products[b]

10 × *TBE buffer*

1 M Tris-borate
20 mM EDTA, pH 8.3
RNA load buffer
    80% formamide, v/v
    0.1% xylene cyanol
    0.1% bromophenol blue
    1 mM EDTA, pH 8.0.

Equipment
Gel apparatus
Power supply

Procedure
1. Prepare *denaturing gel.*[a,b]
2. Denature splicing samples in *RNA load buffer* at 95°C for 30 s and load immediately on gel.
3. Run samples at 50–55°C to prevent formation of RNA secondary structures.
4. The RNA products are visualised by autoradiography.

Notes
a. The polyacrylamide concentration must be optimised for each construct. High percentage acrylamide decreases the mobility of the lariat and intermediate lariat RNA species relatively more than the linear RNA products. It is often convenient to move the lariat species to a position above the precursor RNA by using a sufficiently high polyacrylamide concentration. Constructs with small introns (70–150 bps) are most conveniently run on a 15% gel.
b. The TBE concentration also affects the relative migration of the bands, albeit less pronounced than the polyacrylamide concentration. Lowering the ionic strength (0.5 × TBE) decreases the mobility of the lariat and intermediate lariat RNA species relatively more than the linear RNA products. Generally 0.75 × TBE is a good starting point.

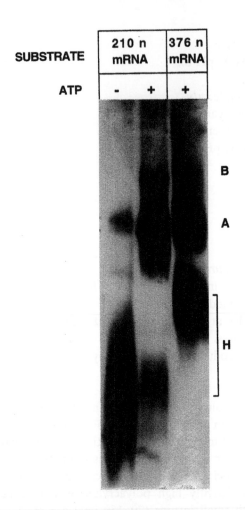

Fig. 5.2. Native polyacrylamide gel analysis of splicing complexes assembled on an intron-containing mRNA. A 210 nucleotide mRNA is incubated under splicing conditions for 20 min in the absence or presence of ATP (lanes 1 and 2) and compared to a 376 nucleotide mRNA treated under the same conditions (lane 3). Three major complexes are formed, the H, A and B complexes. The smallest complex H forms independently of ATP and probably contains variable amounts of hnRNP proteins associated with the RNA. Note that the migration of this complex varies greatly with the length of the RNA. Longer RNAs (above 1000 nucleotides) form H complexes which nearly co-migrate with the A complex. When adding ATP,

5.1.3. Analysis of RNA splicing complexes by native gel electrophoresis

The assembly of the spliceosome is a multi-step process involving sequential binding of snRNP complexes and nuclear proteins to the RNA substrate. The process can be analysed using native gel electrophoresis.[7] To resolve the large RNP complexes (ranging from 10–60 S in size) a low percentage of polyacrylamide with low cross-linking is required. Several alternative gel systems have been described and the choice depends on the size of the substrate. The procedure described below works well for 100–1000 nucleotides splicing constructs. The gels are extremely fragile, so caution should be exercised in gel preparation and handling. If the gel is used for blotting, it is recommended to make a composite polyacrylamide/agarose gel which is less likely to break. Preparation of this type of gel is described by Lamond and Sproat.[8] An example of native gel analysis of splicing complexes is shown in Fig. 5.2.

Materials
RNP complexes: $0.2–1 \times 10^4$ cpm (generally 5 μl of a radioactive standard splicing reaction will suffice for autoradiography)
Heparin 50 mg/ml
Photo-Flo 200 (Sigma P–7417)
5 × Acrylamide mix (80 : 1)
  19.75% acrylamide
  0.25% bis acrylamide in H_2O
20 × Tris-Glycine
  1.0 M Tris base

---

a U1 and U2 snRNP-containing A complex is formed which peaks around 5–15 min. The B complex, which contain all five snRNPs, is generally visible after 10–15 min and accumulate during longer incubation. The B complex can sometimes be resolved into two bands B1 and B2, which may reflect conformational changes in the spliceosome. The position of the A and the B complexes on the gel is relatively insensitive to the length of the mRNA.

1.0 M Glycine
(pH of the mixture will be ~8.8)
APS 10% (freshly made)
TEMED
Whatman 3 MM paper

Equipment
Glass plates 20 × 20 cm
Spacers and comb 1.2 mm thick
Gel apparatus
Power supply 500 volt

Procedure
1. Clean 20 × 20 cm *glass plates* and 1.2 mm thick *spacers* with H₂O and ethanol.
2. Rinse them with a 1 : 200 dilution of *Photo-Flo 200*. Leave them drying in vertical position. Assemble plates.
3. Prepare 100 ml gel solution by mixing:
    20 ml *5 × Acrylamide mix*
    5 ml *20 × Tris-Glycine*
    75 ml H₂O
    800 µl *10% APS*
    80 µl *TEMED*
4. Pour gel and insert *comb* about 1 cm.[a] Leave to polymerise for at least 1 h.[b]
5. Pre-electrophorese gel at room temperature at 250 volt for 1 h before loading.
6. Prepare samples containing *RNP complexes*.
7. Add 1/10 vol of 25 mg/ml *heparin*, mix, spin and continue incubation at 30°C for 10 min.[c]
8. Spin at 12,000 rpm for 1 min and load 5 µl on gel.[d]
9. Run the gel for 3–5 h at 250–300 volt.[e]
10. Carefully remove one glass plate. Pre-cooling it to 4°C makes the gel more stable (NB: Try to let the gel stick to one glass plate).

11. Transfer gel to *Whatman 3 MM paper*, dry and autoradiograph at $-70°C$ with an intensifying screen.

Notes

a. If the comb is inserted too deeply, the pockets are easily damaged when it is removed.

b. Complex gels should be used the day they are poured.

c. Heparin treatment removes more loosely associated protein components of the spliceosome and can be omitted. In this case the splicing complexes will appear larger on the native gel and the electrophoresis time should be increased in order to separate the complexes.

d. The density of the splicing reaction is sufficiently high to load the sample without addition of loading buffer–dye may be added if desired.

e. The gel may get slightly warm, but the temperature must not exceed 30°C.

5.1.4. Debranching of RNA lariats

The nuclear extract and S100 cytoplasmic fraction contain a debranching activity that catalyses the cleavage of the $2'-5'$-phosphodiester bond of the lariat. The observation that lariat RNA species formed *in vitro* are relatively stable in the nuclear extract, probably reflects that the RNA lariats are protected through association with splicing factors. Following deproteination of the RNA the $2'-5'$-linkage at the branch site is readily cleaved both in nuclear and cytoplasmic extracts in the absence of ATP. The debranching procedure is particularly useful for identification of lariat species (see Fig. 5.3). The debranching procedure below is essentially the same as described by Ruskin and Green.[9]

Material

Lariat RNA (deproteinated by phenol extraction)

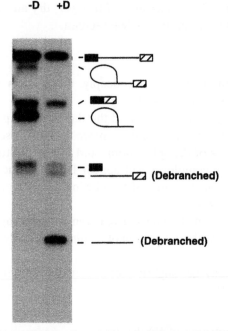

Fig. 5.3. Debranching of lariats. Comparison of splicing products before and after a debranching reaction analysed in a 6% gel containing 1 × TBE. Products obtained from a standard splicing reaction (−D lane) is co-electrophoresed with the same reaction that has been debranched by an additional treatment in S100 extract (+D lane). Identities of individual bands are indicated schematically. Filled boxes, cross-hatched boxes and thin lines denote the 5′- and 3′-exons and intron, respectively. Note that only small amounts of the lariats become debranched under the splicing process, probably due to protection of the branch site by splicing factors. After deproteination, debranching becomes a very efficient process linearising more than 95% of the lariats.

Debranching buffer
  20 mM HEPES-KOH, pH 7.5
  100 mM KCl
  1 mM DTT
  10% glycerol
  8 mM EDTA

Splice stop buffer
  0.5% SDS
  5 mM EDTA
  0.1 M NaCl
  20 mM Tris-HCl, pH 8.0
  0.3 M NaOAc, pH 6.0
  E. coli tRNA, 25 μg/ml
S100 fraction from HeLa cells (Section 3.1.2)

Procedure
1. Dissolve the *lariat* RNA in 20 μl *debranching buffer*.
2. Add 5 μl *S100 fraction*.
3. Incubate for 30 min at 30°C.
4. Cool on ice and add 180 μl of *splice stop buffer*.
5. Extract with 180 μl of phenol/chloroform (1 : 1).
6. Extract with 180 μl chloroform.
7. Precipitate with 2.5 vol EtOH.
8. Redissolve pellet.
9. Analyse by denaturing gel electrophoresis (Section 5.1.2).

5.1.5. Analysis of RNA splicing complexes by sucrose gradients

The classical method to study spliceosome complexes is rate-zonal centrifugation in sucrose gradients. The gradient is prepared as described in Section 4.2.3, and the sample is loaded on the top of the gradient. After centrifugation, 20–30 fractions are collected and analysed. The amount of complexes present in the fractions can be quantified by OD_{260} absorption of the RNA in the complexes and they may be analysed by Northern and/or Western blotting. Alternatively, splicing complexes assembled on an exogenously added RNA substrate can be examined analytically, if the RNA is labelled so quantification of each fraction is possible in a scintillation counter.

Material

Low sucrose buffer
   10% sucrose (w/w)
   10 mM HEPES-KOH, pH 7.6
   60 mM KCl
   3 mM MgCl$_2$
High sucrose buffer
   30% sucrose (w/w)
   10 mM HEPES-KOH, pH 7.6
   60 mM KCl
   3 mM MgCl$_2$

Equipment

Gradient mixer, magnetic stirrer and pump
Ultracentrifuge (Beckman)
SW 40 rotor (Beckman)
5 ml centrifuge tubes

Procedure

1. Precool rotor chamber and rotor to 4°C. Buckets are left on ice until use.
2. The splicing reaction is prepared as described in Section 5.1.1. Store on ice before loading.[a]
3. Sucrose gradients (5 ml per tube) are made from 2.5 ml of each of the *low- and high-sucrose buffers* in a *5-ml centrifuge tube* according to Section 4.2.3 in Chapter 4. Store tubes on ice.
4. The splicing reactions are carefully loaded on top of the gradients placed on ice. Attach buckets to rotor and centrifuge at 4°C at 100,000 g at r$_{av}$ for 2–5 h. [b]
5. Remove the buckets carefully and put them on ice. Remove the tube and collect twenty-four 205-µl fractions from the top of the gradient using a Pippetman and transfer to Eppendorf tubes for further analysis.[c]

Notes

a. If the splicing reactions are not loaded within 1 h, they can be stored at $-80°C$ without any apparent difference in the complex pattern. For analytical investigations of radioactively labelled RNA, a 20 μl splicing reaction will suffice. For preparative purification of splicing complexes up to 200 μl splicing reaction can be loaded on a 5-ml gradient. For larger scales use larger tubes in order to prevent overloading. Loosely associated protein factors can be released by incubating the splicing reaction with heparin prior to loading on the gradient (see Section 5.1.3, Note c). After forming the splicing complexes add 1/10 vol of 50 mg/ml heparin, mix, spin and continue incubation 30°C for 10 min. Centrifuge at 12,000 rpm for 1 min and load supernatant on the gradient.

b. The time and speed of centrifugation may vary depending on the size of mRNA and on the identity of complexes to be resolved. The following parameters are a good starting point to separate pre-spliceosome (complex H and A) and spliceosome (complex B) formed on a 200–500 nucleotide mRNA: A 5–20% sucrose gradient centrifuged at 100,000 g at r_{av} (38,000 rpm in a SW40 rotor) for 5 h can be used with heparin treatment of the splicing reaction, whereas a 10–30% sucrose gradient centrifuged in a SW40 rotor at 38,000 rpm for 2 h can be used without heparin treatment.

c. Sediments may cause problems if fractions are collected from the bottom of the tube.

Comments

The pre-mRNA content of individual fractions may be quantified by scintillation counting (if radioactively labelled) or by Northern blotting. The pre-mRNA content plotted against the fraction number will usually form two peaks (see Fig. 5.4). One slower (lighter) migrating peak containing the A-complex (approximately 40 S) and a faster (heavier) migrating peak containing the fully assembled spliceosome (approximately 60 S in the absence of hep-

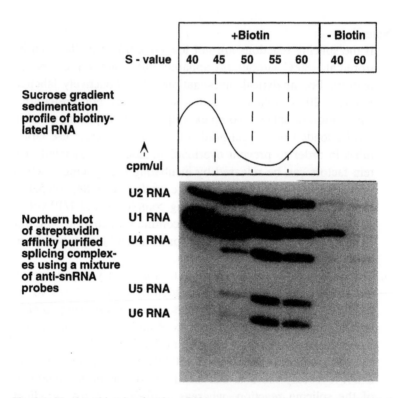

Fig. 5.4. Northern blot showing the snRNA content of splicing complexes separated in a sucrose gradient. Biotinylated and nonbiotinylated pre-mRNA was incubated under splicing conditions and fractionated on a sucrose gradient as described in Procedure. Individual fractions ranging from 35–65S were subjected to affinity chromatography on streptavidin beads under native conditions. The recovered RNA and the associated splicing complexes were denatured and the snRNA content was determined by Northern blot analysis using a mixture of anti- U1, U2, U4, U5 and U6 snRNAs as a probe. The conditions were as described in Section 5.1.6. The four lanes to the left in the Northern blot correspond to pools of fractions 10–12, 14–16, 18–20 and 21–23 of the gradient containing biotinylated RNA, respectively. The two lanes to the right correspond to fractions 10–12 and 21–23 of a gradient containing nonbiotinylated RNA. We often observe some unspecific binding of U1 snRNP to streptavidine beads.

arin). Using the conditions above the 40 and 60 S peaks appear around fractions 10–12 and 18–22, respectively. Characterisation of individual snRNAs in each fraction is possible by Northern blotting, using a mixture of anti-snRNA probes (Section 5.1.6).

5.1.6. Affinity purification and characterisation of spliceosome complexes

Splicing proceeds through a number of intermediate steps which involve formation of the spliceosomes containing snRNP complexes (U1, U2, U4/U6 and U5 snRNPs) and a number of protein factors. The composition of these complexes formed on a particular mRNA can be analysed by an affinity selection procedure described by Grabowski and Sharp[10] and Bindereif and Green.[11]

The principle of this analysis is to incorporate biotin-labelled residues in the RNA during transcription, assemble the RNA-protein complexes, fractionate the complexes according to size, and immobilise the RNA within individual fractions on streptavidin beads (see Fig. 4.3). Finally, the nuclear proteins/RNAs are eluted, identified and quantitated by Northern or Western blotting (Fig. 5.4). Although this method was originally designed to analyse the composition of the spliceosome it can be used as a general affinity purification system.

The following procedure is optimised for separation of splicing complexes in the 20–60 S range and is based on the Grabowski and Sharp procedure[10] with modifications by Kjems and Sharp.[12] For analysis of different sized complexes, adjust the centrifugation time and the sucrose gradient accordingly.

Affinity selection of splicing complexes
Materials
Biotinylated/nonbiotinylated RNA substrate[a]
Streptavidin agarose (BRL)
Nuclear extract (see Section 3.1.1, for HeLa cell extract preparation)

E. coli tRNA (20 mg/ml)[b]
BSA (20 mg/ml)
KCl (3 M)
2 × splicing buffer
  2 mM ATP
  10 mM creatine phosphate
  5 mM $MgCl_2$
  0.6 mM dithiothreitol
  0.5 μg/μl *E. coli* tRNA[b] (optional – see Note e in Section 5.1.1)
RNasin (5 units/μl)
NET 2 buffer
  50 mM Tris-HCl, pH 7.9
  0.05% Nonidet P-40 (v/v)
  0.5 mM DTT
Formamide loading buffer
  80% formamide
  0.05% xylene cyanol
  0.05% bromphenol blue
  1 mM EDTA, pH 8.0
Low sucrose buffer
  10% sucrose (w/w)
  10 mM HEPES-KOH, pH 7.6,
  60 mM KCl,
  3 mM $MgCl_2$
High sucrose buffer
  30% sucrose (w/w)
  10 mM HEPES-KOH, pH 7.6
  60 mM KCl
  3 mM $MgCl_2$
KCl (3 M)
Washing buffer
  0.05% Nonidet P-40 (v/v)
  500 mM KCl
  20 mM HEPES-KOH, pH 7.6
  1 mM $MgCl_2$

0.5 mM DTT
0.1 mM EDTA
Elution buffer
  20 mM Tris-HCl, pH 7.5
  4 M Urea
  0.5% SDS
  10 mM EDTA
  0.3 M NaCl
5 × SDS load buffer
  25% glycerol
  10% SDS
  250 mM Tris-HCl, pH 6.8
  0.025% (w/v) Commassie Blue
  5% 2-mercaptoethanol

Equipment
Ultracentrifuge, SW 55, 5 ml tubes
Gradient mixer, magnetic stirrer and pump
Scintillation-counter, vials and -liquid
Rotating wheel (or rocking table)
Water baths/heating blocks (30, 37 and 95°C)

Procedure
1. Before assembling the RNA-protein complexes make sure that the ultracentrifuge(s) is set at 4°C and produce the desired number of 5 ml 10–30% sucrose gradients (see Section 4.2.3 for preparation). Balance opposite tubes when placed in the buckets with low-sucrose buffer. Leave 3–4 mm free space on top of the gradient for the sample. Make sure that you make the gradients exactly the same way and leave them on ice until use.
2. Assemble RNA-protein complexes in 100 µl reactions under splicing conditions by mixing:

    1 µg of *biotinylated* or *nonbiotinylated RNA substrate* in H_2O

50 μl of freshly made *2 × splicing buffer*
30 μl of pre-cleared *nuclear extract*[c]
Adjust volume to 100 μl with water.

3. Mix by flicking the tube, and incubate samples for 20 min at 30°C

4. Carefully load the whole reaction on a 5-ml 10–30% sucrose gradient and store the tube in the swinging bucket on ice.

5. Centrifuge the gradients for 2 h and 50 min at 48,000 rpm in a *SW 55 rotor* at 4°C (brakes off).[d]

6. During centrifugation, pre-block the *streptavidine beads*.[e]

7. Gently collect 24 fractions of 205 μl from the top of the gradient using a P200 pipette.

8. Transfer 25 μl of each of the fractions to *scintillation vials*. Add 5 ml of *scintillation liquid* to the vial and count in a *scintillation counter* to determine the amount of labelled mRNA in each fraction. Plot the counts (see Fig. 5.4).

9. Pool 2–3 fractions around the peaks.

10. Adjust the salt concentration to 300 mM KCl by adding 0.08 vol of 3 M *KCl* (13 μl for each fraction that is pooled; inject slowly while mixing gently).

11. Add 30 μl of a 50% suspension of pre-blocked *streptavidine beads* in *NET 2 buffer* to each pool and rotate at 4°C for 1 h.[e]

12. Wash beads 4 times with 1 ml *washing buffer* over a total of 2 h.

If complexed nuclear RNA is to be analysed follow steps 13–15. If complexed protein is to be analysed, follow step 13b
Analysis of complexed nuclear RNA

13. Release complexed RNAs by incubating collected beads in 300 μl *elution buffer* at 50°C for 5 min.

14. The supernatant is extracted once with 1 vol of phenol/chloroform, once with chloroform.

15. Add 5 μg of *E. coli tRNA* and precipitate by adding 2.5 vol 96% EtOH, wash pellet in 80% EtOH, dry and resuspend in

formamide loading buffer. The RNA is analysed by Northern blotting as described below.

Analysis of complexed protein

13b. Release complexed protein by incubating the beads at 95°C for 1 min in 30 μl 1 × *SDS load buffer* and analyse by Western blotting.

Notes

a. Preparation of biotinylated RNA transcripts is described in Section 2.3.2. The biotin is introduced randomly into the RNA transcript by including biotinylated nucleotides during the transcription. Preferably, only a few positions in the transcript should be biotinylated. More excessive biotinylation may interfere with complex assembly and increase the amount of RNA remaining in the phenol phase during extraction. We find that including 10% biotin-16-UTP of the total UTP concentration during the transcription works for 200–500 nucleotide transcripts (T7 RNA polymerase nucleotide usage is biased towards the nonbiotinylated species). The ratio should be increased to 20–30% for shorter RNAs (see Grabowski and Sharp[10] for further discussion of this issue). Radio-labelled nucleotides may be included to ease subsequent quantification. ^{35}S or ^{14}C labelled nucleotides are recommended to eliminate interference with subsequent analysis. Capping of the transcript by including GpppG during the transcription (Section 2.3.2) is desirable but may be omitted for economic reasons, since it is costly to include the capped nucleotide in preparative RNA synthesis. Use UV-shadowing to locate the transcript in the gel (Section 2.1.2). The concentration of the RNA is calculated from the specific activity of incorporated ^{35}S label. A transcription volume of 100 μl generally yields 10–50 μg of biotinylated RNA which will suffice for many purposes. As a control for unspecific binding to the strepavidine beads in the subsequent affinity purification experiment, it is recommended to perform a nonbiotinylated RNA transcription in parallel.

b. The tRNA used as carrier RNA must be RNase free (use *E. coli* MRE 600 tRNA from Boehringer Mannheim). The tRNA is phenol-extracted, precipitated, and resuspended in water at a concentration of 20 mg/ml.

c. Clear the nuclear extract for insoluble material just before use by centrifuging 1 min at 14,000 rpm.

d. Under these conditions the 60 S spliceosome migrates around fraction 20 and 30 S around fraction 10.

e. Pre-blocking of streptavidine-agarose beads: (i) wash beads twice with a 10-fold volume of *NET 2 buffer*. Centrifuge beads for 30 s at 2000 rpm between washes; and (ii) rotate beads with a 10-fold volume of *NET 2 buffer* containing 0.5 mg/ml *E. coli* tRNA and 0.2 mg/ml *BSA* in *NET* 2 at 4°C for 1 h. Recover beads and add 1 vol. of NET 2 buffer. Store on ice. Magnetic beads may also be used but we have experienced problems with RNase contamination and higher background in nuclear extracts.

Northern blot analysis of mixed RNAs using multiple snRNA probes

Recovered RNA may be analysed by 3'-end labelling with pCp (Section 2.3.3.2) followed by denaturing gel electrophoresis. Breakdown products due to water hydrolysis or contaminating RNases have a 3'-phosphate and will not be 3'-end labelled by pCp (see e.g. Ref. 11). However, if known species of RNA are to be analysed, RNase protection or Northern blotting are the recommended methods. The procedure below describes the quantification of the 5 snRNAs in spliceosomal complexes (U1, U2, U4, U5, and U6 snRNAs), essentially as described by Grabowsky and Sharp.[10] The RNAs are separated in a denaturing polyacrylamide gel, transferred to a membrane and probed with a mixture of five antisense snRNA probes (Fig. 5.4).

Materials
Zeta-Probe GT filters (Biorad)
Nuclear extract
Anti-sense U1-, U2-, U4-, U5- and *U6-snRNP riboprobes*
TBE buffer (10 ×)
   1 M Tris-Borate, pH 8.3
   10 mM EDTA
Polyacrylamide gel, 20 × 20 cm, 0.5 mm thick, 8%, 8 M Urea,
0.5 × TBE gel

Equipment
Gel apparatus
Electrophoretic transfer cell unit (Biorad)
Power supply 500 volt
Magnetic stirrer
Vacuum oven
Sealing unit and *plastic bags*
Incubator for hybridisation

Method
1. Fill *electrophoretic transfer cell unit* with 2–3 litres of *0.5 × TBE*
   and place on *magnetic stirrer* in the cold room (to be used in
   step 5).
2. Prepare a *polyacrylamide gel*, and pre-run the gel for 30 min.
3. Denature samples by incubating them at 95°C for 30 s and load
   on gel.
4. Run gel hot (15 W) until the xylene cyanol dye is 2/3 down.
5. The RNA is transferred to Zeta-Probe GT filters (or equivalent).[a]
6. Rinse filter in *0.5 × TBE* and bake in a *vacuum oven* at 80°C
   for 30 min.[a]
7. Store filter in a sealed *plastic bag* at room temperature.
8. Probe the filter with *anti-sense transcripts* of U1, U2, U4, U5
   and U6 snRNAs using the formamide procedure given by the
   manufacturer.[b]

9. Gels are autoradiographed (see Fig. 5.4) and quantified using a PhosphorImager.[c]

Notes

a. Conditions may depend on which type of membrane is used. Follow instructions given by the manufacturer. Transfer of small RNAs is efficient by electroblotting at 2 V/cm for 1 h followed by 5 V/cm for 2 h (Keep the effect below 10 W).

b. The anti-snRNA probes for U1, U2, U4, U5, and U6 snRNAs were prepared according to Konarska and Sharp.[7] To obtain equal U1, U2, U4, U5 and U6 snRNA signals in the 60 S spliceosome peak the corresponding antisense probes should be mixed in the following ratios of radioactivity: 1 : 2 : 4 : 2 : 2, respectively.

c. The exact quantification of U4 snRNP may require re-hybridisation of the filter with anti-U4 snRNA probe alone due to the frequently observed co-migration of a degradation product of U1 snRNA with U4 snRNA (Fig. 5.4).

5.2. Polyadenylation

5.2.1. In vitro *polyadenylation*

The 3'-end of an eukaryotic mRNA is formed post-transcriptionally by endonucleolytic cleavage downstream of the coding sequence followed by extension of the upstream fragment by approximately 200 adenylate residues. In higher eukaryotes the specificity of the reaction depends on two conserved sequences in the pre-mRNA; a highly conserved AAUAAA 10–30 nucleotides upstream from the cleavage site and a G/U-rich sequence within 50 nucleotides downstream from this site. The reaction requires a complex set of nuclear proteins, and a reconstituted polyadenylation reaction from pure proteins has not yet been achieved. However, processing of the 3'-end has been intensively studied using partially purified fractions, and for many purposes polyadenylation can be studied using a crude nuclear extract. The protocol for *in vitro* polyadenylation was originally designed by Moore and Sharp[13] and a more detailed protocol has been published by Keller and colleagues.[14]

Materials
Buffer D
    20 mM HEPES-KOH pH 7.9
    1.5 mM $MgCl_2$
    100 mM KCl
    0.2 mM EDTA
    0.5 mM DTT
    20% glycerol
Proteinase K buffer
    10 mM Tris-HCl, pH 7.5
    10 mM EDTA
    100 mM NaCl
    0.5% SDS
Formamide load buffer
    80% (v/v) formamide
    0.1% xylene cyanol
    0.1% bromophenol blue
    1 mM EDTA, pH 8.0
ATP, 25 mM
Creatin phosphate, 0.5 M (Sigma)
E. coli tRNA, 1 µg/µl (Boehringer Mannheim)
Polyvinylalcohol, 10% (Sigma)
Recombinant RNasin (Promega)
RNA substrate (1–50 fmol; 1000–10 000 cpm/fmol)[a]
HeLa cell nuclear extract in buffer D[b]
Proteinase K, 10 mg/ml (Merck)

Procedure
1. For a polyadenylation reaction in 25 µl add to a tube:

4.25 µl H$_2$O
1 µl 25 mM *ATP*
1 µl 0.5 M *creatine phosphate*
1 µl 1µg/µl *tRNA*
6.25 µl 10% *polyvinylalcohol*
0.5 µl *RNasin*
RNA substrate (1–50 fmol/reaction)
11 µl *HeLa cell nuclear extract*[c]

2. Mix by pipetting and incubate the reaction at 30°C for 30 min. Take out 5–10 µl of the reaction for analysis of complex formation in native gels, if desired.[d]

3. Stop the reaction by adding 180 µl of *proteinase K buffer* supplemented with 5 µg *E. coli tRNA* and 50 µg *Proteinase K*. Incubate at 30°C for 10 min.

4. Extract once with phenol/chloroform and recover the aqueous phase.

5. Extract once more with chloroform and precipitate with ethanol (see Section 2.1.2).

6. Resuspend the pellet in 5–10 µl *formamide load buffer*.

7. Denature the RNA by incubation at 95°C for 2 min and chill on ice.

8. Load on a 5–8% denaturing polyacrylamide gel.

9. Detect RNA by autoradiography.

Notes

a. A widely used model RNA substrate for *in vitro* polyadenylation is the adenovirus L3 polyadenylation site.[13] The RNA is generally synthesised by *in vitro* transcription (Section 2.3.1). The 5'-cap is not essential but greatly stabilises the RNA in the nuclear extract, and [α-^{32}P]UTP is usually included so the formation of product can be followed by autoradiography.

b. A slight modification of the reaction conditions allows *in vitro* polyadenylation using extracts prepared from lymphoid cells.[15]

c. The HeLa cell nuclear extract or polyadenylation factors diluted with buffer D to a final volume of 11 µl. The amount of nuclear extract has to be determined empirically. Nuclear extract is made in buffer D as described in Section 3.1 (Chapter 3). The final concentrations contributed by buffer D are 8.8 mM HEPES, pH 7.9, 8.8% glycerol, 44 mM KCl, 0.088 mM EDTA, 0.66 mM MgCl$_2$, 0.22 mM DTT.

d. The reaction is stopped by adding heparin to 0.2 mg/ml final concentration followed by 10 min continued incubation at 20°C. Native gel electrophoresis is performed as described for the analysis of splicing complexes (Section 5.1.3).

5.2.2. Analysis of mRNA polyadenylation states

Polyadenylation states are coupled to developmental events reflecting recruitment of stored cytoplasmic mRNAs into translationally active species, and certain mRNA turnover pathways are initiated by deadenylation. Moreover, recent work with yeast has suggested that there is an intimate connection between the poly(A)-tail and translation initiation via the poly(A)-binding protein, probably mediated by the eIF–4G factor. Therefore, the ability to measure dynamic changes of poly(A)-tail length is an important tool both in developmental studies and in general translational control. Hitherto, poly(A)-tail length has been estimated by H-blotting, involving cleavage by RNase H followed by Northern analysis. The major drawbacks of the H-blotting procedure are the modest sensitivity and the poor resolution. Recently Strickland and colleagues have developed a PCR-based poly(A) test that is both sensitive and exhibits a good resolution[16] (Fig. 5.5). Importantly, it is possible to carry out the analysis in a single tube, and the produced cDNA can be used for the analysis of multiple mRNAs since the target is defined by the upstream PCR primer.

Materials

5 × superscript RT buffer
   250 mM Tris-HCl, pH 8.3
   375 mM KCl

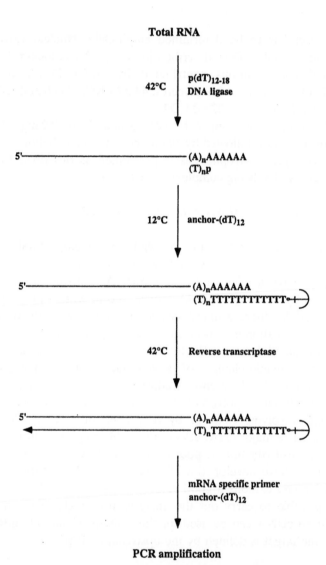

PCR amplification

Fig. 5.5. Outline of a PCR-based procedure for measuring the length of the poly(A) tail. Total RNA is hybridised with a saturating amount of phosphorylated olig-o(dT)$_{12-18}$ at 42°C in the presence of T4 DNA ligase resulting in a complementary poly(dT) tract except at the 3'-terminus of the mRNA. The temperature is then

15 mM MgCl$_2$
Phosphorylated $(dT)_{12-18}$
DTT (10 mM)
dNTPs (10 mM)
ATP (10 mM)
T4 DNA ligase (10 Weiss units/µl) — USB
Oligo(dT)-anchor : 5'-dGCGAGCTCCGCGGCCGCGT$_{12}$ (200 ng/µl)
Superscript H⁻ reverse transcriptase (400 units/µl) — BRL

Method

1. Mix 20 ng phosphorylated (dT)$_{12-18}$ and 10 ng total RNA in 7 µl H$_2$O and denature at 65°C for 5 min. Transfer sample to 42°C.

2. Add 13 µl of the following mixture prewarmed to 42°C:
   4 µl H$_2$O
   4 µl *5 × superscript RT buffer*
   2 µl 10 mM *dithiothreitol*
   1 µl 10 mM *dNTPs*
   1 µl 10 mM *ATP*
   1 µl 10 U/µl *T4 DNA ligase*
   Incubate for 30 min at 42°C.

3. Add 1 µl *oligo(dT)-anchor* (200 ng/µl) at 42°C.

4. Mix and incubate at 12°C for 2 h.[a]

5. Transfer the sample to 42°C for 2 min.

6. Add 1 µl *superscript H⁻ reverse transcriptase* and incubate at 42°C for 1 h.

---

reduced to 12°C in the presence of a fivefold molar excess of an oligo(dT)-anchor primer that will preferentially anneal to the 3'-terminus of the mRNA and be ligated to the nascent poly(dT) strand. The temperature is now raised to 42°C, and reverse transcriptase generates cDNA using the poly(dT)-anchor as primer. At this stage the poly(A) test cDNA library is ready, and the polyadenylation status of an individual mRNA can be measured by PCR using a mRNA-specific upstream primer and the oligo(dT)-anchor primer as the downstream partner.

7. Inactivate the enzymes at 70°C for 30 min.
8. Use 1 µl cDNAs for a 50 µl PCR.[b]

Notes

a. 12°C for 2 h is easily maintained by tap-water in a polystyrene box.
b. The poly(A) test cDNA library should be examined by a standard PCR using the upstream target primer and a downstream specific primer, before the cDNA library is subjected to the poly(A) analysis to ensure that the desired mRNA is represented as cDNA in the library. Use 25 pmol of target primer and 25 pmol of oligo(dT)-anchor primer and 62°C as the annealing temperature in the poly(A) analysis proper, and use nested priming to establish specificity. The PCR products can be visualised by agarose gel electrophoresis and ethidium bromide staining, or alternatively tracer amounts of [α-^{32}P]dNTP can be included during PCR and the products analysed by polyacrylamide gel electrophoresis. The latter approach provides higher sensitivity and resolution. An important control of PCR specificity is restriction enzyme cleavage that should generate a constant 5'-fragment regardless of physiological changes of the polyadenylation states.

5.3. RNA modifications

RNA transcripts are prone to post-transcriptional processes leading to changes in the primary sequence of the RNA molecules. These changes are often referred to as *editing* if the base is changed into another canonical base (A, U, C or G(I)) or *modifications* if the nucleotide is changed into a noncanonical base. There are two types of editing; insertion/deletion of nucleotides or substitution/conversion of nucleotides. The use of the terms RNA modifications and the substitution/conversion type of RNA editing is ambiguous.

Some investigators limit the term editing to alterations in the protein-coding region (excluding splicing) of mRNAs that lead to changes in the protein encoded by the mRNA. Other investigators use editing to describe changes in RNA transcripts independent of whether it occurs in protein-coding or untranslated regions of the RNA, and restrict the term RNA modification to alterations in stable RNAs (rRNA, tRNA, snRNA etc). From a chemical point of view the conversion type of RNA editing is a subtype of RNA modifications.

Different types of substitution/conversion-editing have been described. The best characterised involves cytidine deamination to uridine in the editing of transcripts for apolipoprotein B, and adenosine deamination to inosine in the editing of the transcripts encoding the glutamate receptor subunits. More than 90 different types of base modifications have been described. Moreover, methylations and substitutions of the ribose and phosphate moieties have also been identified. Recently, the 2'-O-ribose methylation has been the focus of much attention since the discovery that the snoRNA guides the methylation in a *trans*-acting fashion.

The assay of choice depends on the RNA change to be investigated and can be classified as follows:
1. Editing (base change).
2. RNA modifications that interfere with reverse transcription (e.g. modifications at base-pairing positions or bulky groups).
3. RNA modifications which do not directly interfere with reverse transcription (e.g. 2'-OH methylations).

Editing can be studied with small quantities of RNA since a RT-PCR step is included to amplify the RNA. However, studying RNA modifications requires direct analysis of the RNA. Different methods, including HPLC, TLC and mass spectroscopy, have been employed to identify and quantify the modifications. A description of these methods is beyond the scope of this book.

However, quantification of the extent of editing or modifications can often be assayed by a primer extension assay. The principle of a primer extension analysis is illustrated in Fig. 5.6; that shows

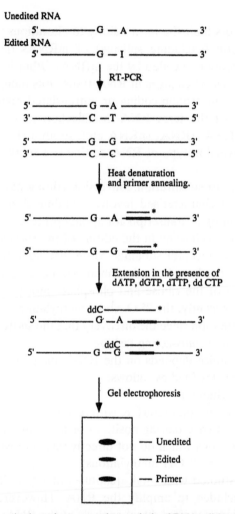

Fig. 5.6. Quantitative primer extension analysis of RNA editing after RT-PCR. The illustration shows an A to inosine (I) conversion, but the method can be used for any base change. The RT-PCR will convert the inosine to a G. A ^{32}P end-labelled (asterisk) primer is annealed 3′ of the edited residue at a distance without any G's between the 3′-end of the primer and the edited site. Extension in the presence of ddCTP, dATP, dTTP and dGTP produces two bands depending on whether the transcript is edited or nonedited.

the analysis of an A to G conversion. In this case, the primer is complementary to a sequence 3′ of the edited nucleotide and placed at the maximum distance from the edited site, dictated by the absence of G's in the RNA template between the primer site and the edited base. Extension of the primer with reverse transcriptase (or Klenow polymerase if the analysis is performed on a PCR product) in the presence of dideoxyCTP, dATP, dTTP and dGTP will proceed until the first G on the template is reached. Thus, for the edited template the extension will terminate at the introduced G, while for the unedited template the extension will continue to the following G. The quantities of edited versus unedited transcripts can then be determined after an electrophoretic separation of the cDNA bands.

Similar assays can be used to study modifications that inhibit the extension of the reverse transcriptase directly and also modifications which render the RNA susceptible to selective strand scission at the modified residue (e.g. m7 guanosines, Section 4.4.2.2). To quantify the extent of modification, one of the nucleotides in the extension mixture should only be present in the dideoxy form (Fig. 5.7). The primer extension assay can also be used for some modifications which do not affect the extension, provided the modification interferes with the reactivity of the nucleotide. The latter approach has been used to study the 2′-O-methylations of guanosine where the modification renders the modified guanosine resistant to RNase T1 digestion. Introduction of a complete RNase T1 digest before the primer extension allows the detection of the modification (Fig. 5.8a). However, the approach requires that there is no nucleotide as the one modified for at least 12 nucleotides 3′ of the modified yresidue. If this is not the case alternative approaches can sometimes be employed (see Fig. 5.8b).

5.3.1. Reverse transcriptase–polymerase chain reaction (RT-PCR)

Materials
10 × extension buffer

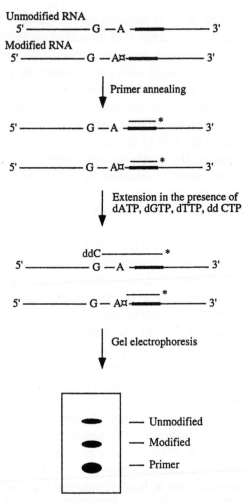

Fig. 5.7. Primer extension of a modification which abolish reverse transcriptase extension. The extension is performed on the RNA. A ^{32}P end-labelled (asterisk) primer is annealed 3' of the residue modified at a distance without any G's between the 3'-end of the primer and the modified site. Extension in the presence of ddCTP, dATP, dTTP and dGTP produce two bands which can be used for determination of the degree of modification.

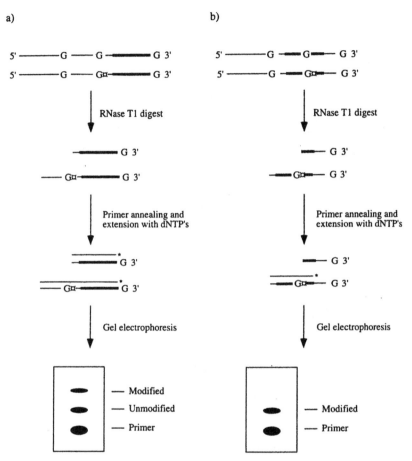

Fig. 5.8. Primer extension analysis of a modification that does not affect reverse transcriptase extension directly. (a) 2′-O-methylation, which renders the G insensitive to RNase T1 digestion, is used as an example. Analysis of a methylated G where there are no G's for 12 nucleotides 3′ of the methylated G. (b) Analysis of a G modification in a sequence without other G's for at least 12 nucleotides around the modified G. The latter situation only detects the presence of the modification but does not allow quantification.

100 mM Tris-HCl, pH 8.3
50 mM MgCl$_2$
500 mM KCl
10 mM DTT
PCR buffer
   10 mM Tris-HCl, pH 8.3
   1.5 mM MgCl$_2$
   100 mM KCl
   0.001% gelatin
RNasin (Promega)
Moloney murine virus reverse transcriptase (superscript II; GIB-CO/BRL)
dNTPs 10 mM each
Taq polymerase

Equipment
PCR machine

Procedure
1. Make 10 μl solution containing:
   H$_2$O to 10 μl to final volume of 10 μm
     0.25 μg total *RNA*
     0.2 unit *RNasin*
     50 units *Moloney murine virus reverse transcriptase* (super-script II)
     1 μl *10 × extension buffer*
     1 μl *10 mM dNTP solution*
2. Incubate 10 min at RT followed by 50 min at 37–42°C.[a]
3. Add 40 μl PCR buffer containing 50 pmol of the appropriate primers and *2 units taq polymerase.*
4. Amplify the template using standard PCR methods (35 rounds).
5. Load the PCR product on an agarose gel and purify the appropriate band.[b]

Notes

a. If the cDNA reaction (after step 2) is divided into aliquots for different PCR reactions dNTP's should be added in step 3 to a final concentration of at least 0.15 mM in the PCR reaction.

b. If low temperature melting agarose is used for the electrophoresis subsequent primer extension can be performed directly in the melted gel.

DNase treatment

For most purposes, contamination from genomic DNA can be eliminated in the RT-PCR reaction by designing primers that span an intron. The size difference between PCR products generated from genomic DNA and mRNA will preclude contamination.

If necessary residual DNA can be removed by a DNase treatment.

Materials

DNase buffer
   100 mM NaOAc, pH 5.0
   10 mM MgSO₄

RNase-free DNase (Boehringer Mannheim)
RNasin (Promega)
NaOAc 0.4 M (pH 5.0)

Procedure

1. Incubate 5 μg total RNA for 1 h at 37°C with 10 units *RNase-free DNase* and 1 unit *RNasin* in 50 μl *DNase buffer*.
2. Add 50 μl 0.4 M NaOAc (pH 5.0).
3. Phenol/chloroform extract.
4. Chloroform extract.
5. Add 250 μl EtOH (leave on dry ice for 15 min) and precipitate.

5.3.2. Primer extension by Klenow polymerase (after RT-PCR; Fig. 5.6)

Material

10 × annealing buffer
   100 mM Tris-HCl, pH 7.8
   50 mM $MgCl_2$
   500 mM KCl
Klenow X-dideoxynucleotide mix
   0.1 mM dNTP (-dXTP)
   1 mM ddXTP
Formamide load buffer
   80% formamide
   1 mM EDTA, pH 8.00
   0.02% xylene cyanol
   0.02% bromophenol blue
5'-end-labelled extension primer
Klenow polymerase

Procedure

1. Denature the DNA fragment[a] (5.3.1) in 5 μl H_2O by incubating at 100°C for 5 min, followed by a quick cooling on ice.
2. Add 4 μl *5'-end-labelled extension primer* and 1 μl *10 × annealing buffer*.
3. Incubate 15 min at 37°C.
4. Add 5 μl *Klenow X-dideoxynucleotide mix* and 1 unit *Klenow polymerase*.
5. Incubate for 10 min at 37°C.
6. Terminate the reaction by addition of 5 μl *formamide load buffer*.
7. Heat to 95°C for 3 min and chill on ice prior to loading on a 15 or 20% denaturing polyacrylamide gel.[b]

Notes

a. An electrophoretic purification of the PCR product is necessary to remove residual nucleotides. Furthermore, the size selection might eliminate contaminations due to artificial products generated during the PCR.

b. The editing efficiency can be calculated either by excising the bands from the gel or using a PhosphorImager system.

5.3.3. Primer extension by reverse transcriptase (Fig. 5.7)

Materials

10 × annealing buffer
   100 mM Tris-HCl, pH 7.4
   400 mM KCl
   5 mM EDTA
25 × RT buffer
   1.25 M Tris-HCl, pH 8.0
   250 mM $MgCl_2$
   50 mM DTT
Maxam–Gilbert load buffer
   80% deionised formamide
   10 mM NaOH
   1 mM EDTA
   0.02% xylene cyanol
   0.02% bromophenol blue
RT-X-dideoxynucleotide mix
   2.5 mM (dNTP-dXTP)
   5 mM ddXTP
dNTP 2.5 mM each
AMV reverse transcriptase (Life Science)
tRNA carrier 5 mg/ml
5-end-labelled extension primer
NaOAc 0.3 M (pH 6.0)

Procedure
   1. Mix ⩾0.5 fmol RNA and approximately 10 fmol ^{32}P *5'-end-*

    labelled extension primer in 6 µl annealing buffer (including *0.1 µg/µl tRNA carrier*).[a]

2. Heat to 95°C for 30 s followed by a 30-min incubation at 48°C (for primers ≤18 nucleotides anneal at 37°C).
3. Spin down evaporated drops.
4. Add 4 µl extension mixture (made of 0.4 µl *25 × RT buffer*, 0.8 µl *RT-X-dideoxy-nucleotide mix*, 2.8 µl H$_2$O and 1 unit *AMV reverse transcriptase*).
5. Incubate for 30 min at 37°C.
6. Terminate the reaction by addition of 40 µl 0.3 M *NaOAc* (pH 6.0) and precipitate with 125 µl EtOH.
7. Wash the pellet in 70% EtOH.
8. Redissolve the dried pellet in *Maxam–Gilbert load buffer*.
9. Heat the samples to 95°C for 3 min and chill on ice.
10. Load on a 20% denaturing polyacrylamide gel.[b]

Notes
a. The applied carrier RNA should not contain sequences complementary to the extension primer.
b. The editing efficiency can be calculated either by excising the bands from the gel or using a PhosphorImager system.

5.4. Translation

5.4.1. Preparation of ribosomes

The ribosome provides an easily accessible source of endogenous RNA-protein complexes that participate in the overall process of translation. In this section, we present protocols for obtaining salt-washed ribosomes from *E. coli* and mammalian cell-lines. However, if highly active 'vacant couples' devoid of tRNAs and mRNAs is the aim, it is necessary to use more elaborate protocols that include dissociation of bacterial 70 S 'tight couples' into subunits,[17] and puromycin-stripping of eukaryotic 80 S ribosomes followed by high-salt dissociation into subunits.[18] The important difference between bacterial and eukaryotic ribosomes is their Mg^{2+} sensitivity. Bacterial ribosomes can be released into subunits by lowering the Mg^{2+} concentration to 1 mM, whereas eukaryotic ribosomes cannot be dissociated into subunits simply by lowering the Mg^{2+} concentration.

5.4.1.1. Bacterial ribosomes
Materials
Extraction buffer
  20 mM Tris-HCl, pH 7.6
  10 mM $MgCl_2$
  100 mM NH_4Cl
  0.5 mM EDTA
  6 mM 2-mercaptoethanol
Sucrose cushion
  1.1 M sucrose
  20 mM Tris-HCl, pH 7.6
  10 mM $MgCl_2$
  500 mM NH_4Cl
  0.5 mM EDTA
  6 mM 2-mercaptoethanol
High-salt buffer
  20 mM Tris-HCl, pH 7.6
  10 mM $MgCl_2$
  500 mM NH_4Cl
  0.5 mM EDTA
  6 mM 2-mercaptoethanol
Storage buffer
  10 mM Tris-HCl, pH 7.6
  10 mM $MgCl_2$
  60 mM NH_4Cl
  3 mM 2-mercaptoethanol

Alumina, e.g. Sigma's type A-5
DNase I (10 mg/ml)

Equipment
Mortar—prechilled at −20°C
Ultracentrifuge
Fixed-angle rotor (e.g. Beckman type 35)

Procedure (to be carried out at 0–4°C)
1. Grind 25 g frozen *E. coli* MRE600 cells[a] with 50 g *alumina* and about 1 ml *extraction buffer* in a mortar until a sticky 'popping' paste is obtained.
2. Gradually add 45 ml of *extraction buffer* and continue grinding.
3. Add 10 μl *DNase I* and grind until a reduction in viscosity is observed.
4. Centrifuge at 30,000 g for 45 min to remove debris and alumina.
5. Layer the supernatant[b] on two 35 ml *sucrose cushions* in 75 ml polycarbonate tubes.
6. Centrifuge at 95 000 g for 22 h (35,000 rpm in a Beckman type 35 rotor).
7. Rinse the pellet twice with 10 ml of *extraction buffer*.
8. Resuspend the two pellets in a total volume of 5 ml of *extraction buffer*.[c]
9. Centrifuge at 27,000 g for 15 min at 4°C (15,000 rpm in a Sorvall SS34 rotor) to clarify the resuspended solution.
10. Remove the supernatant and increase the volume to 100 ml with *high-salt buffer*.
11. Pellet ribosomes by ultracentrifugation for 7 h at 38,000 g (22,000 rpm in a Beckman type 35 rotor).
12. Rinse the pellet twice with 10 ml of *extraction buffer*.
13. Resuspend the pellet in *storage buffer*[c] at a concentration of about 500 A_{260}/ml.[d]
14. Clarify the salt-washed ribosomes by centrifugation at 27,000 g for 15 min.
15. Divide into small aliquots, quick-freeze, and store at −80°C.

Notes

a. Cells are harvested at late logarithmic growth and chilled slowly to allow ribosome run-off. They are washed with 20 mM Tris-HCl, pH 7.6, 10 mM magnesium acetate, and kept frozen at $-80°C$.

b. Often called the S30 supernatant.

c. Place a small magnetic stirring bar in the ultracentrifuge tube and stir slowly for about 1 h on ice.

d. One A_{260} unit corresponds to 23 pmol of 70 S ribosomes, but recall that 2-mercaptoethanol absorbs at 260 nm (Section 2.1).

5.4.1.2. Mammalian ribosomes

Materials

Phosphate-buffered saline
   10 mM NaH_2PO_4
   30 mM K_2HPO_4
   125 mM NaCl
   pH 7.4

Lysis solution
   20 mM Tris-HCl, pH 8.5
   1.5 mM $MgCl_2$
   140 mM KCl
   1 mM dithiothreitol
   0.5% Nonidet P-40

Sucrose cushion
   30% (w/v) sucrose
   35 mM Tris-HCl, pH 7.8
   10 mM $MgCl_2$
   600 mM KCl
   1 mM dithiothreitol

Storage buffer
   20 mM Tris-HCl, pH 7.6
   5 mM $MgCl_2$
   100 mM KCl
   1 mM dithiothreitol

0.25 M sucrose
KCl (2 M)
MgCl₂ (1 M)
Sodium deoxycholate (10% — freshly prepared)

Equipment
Ultracentrifuge
Fixed-angle rotor (e.g. Beckman type 70 Ti)

Procedure (to be carried out at 0–4°C)
1. Wash 10^8 cells (about 1 g) with *phosphate-buffered saline*. Following centrifugation, cell pellets may be stored at −80°C.[a]
2. Lyse cell pellets in 5 ml ice-cold *lysis solution* by gently pipetting up and down a few times.
3. Centrifuge at 13,000 *g* for 30 min.
4. Isolate the top two-thirds of the supernatant.
5. Adjust the concentrations of KCl and $MgCl_2$ to 550 mM and 5 mM, respectively, and add 1/10 vol of *sodium deoxycholate*. Mix for a few minutes.[b]
6. Layer the post-mitochondrial supernatant over a 3.5 ml *sucrose cushion* in a 10 ml polycarbonate bottle.
7. Centrifuge at 176,000 *g* for 2 h (50,000 rpm in a Beckman type 70.1 Ti rotor).
8. Discard the supernatant and rinse the pellet with 1 ml *lysis buffer*.
9. Resuspend in storage buffer[c] at a concentration of about 300 A_{260}/ml.[d]
10. Clarify the resuspended ribosomes by centrifugation at 13,000 *g* for 15 min.
11. Divide into aliquots, quick-freeze, and store at −80°C.

Notes
a. Cells in suspension culture are pelleted directly, whereas adherent cells are released by 2 g/l EDTA in *phosphate-buffered*

saline at 37°C. The latter should be washed thoroughly in *phosphate-buffered saline* to minimise carry-over of EDTA.

b. Treatment with sodium deoxycholate ensures release of ribosomes from microsomes

c. Place a small magnetic stirring bar in the ultracentrifuge tube and stir slowly for about 1 h on ice.

d. One A_{260} unit corresponds to 18 pmol of 80 S ribosomes, but recall that dithiothreitol absorbs at 260 nm (see Section 2.1, Chapter 2).

5.4.2. Preparation and analysis of polysomes

In both pro- and eukaryotic organisms actively translated mRNAs are present in polyribosomes, whereas stored or masked mRNAs are present in mRNPs. In studies of translational control, it is desirable to be able to discriminate between these two populations of endogenous mRNAs. In general, polysomal mRNA particles are larger than mRNP particles, so the two populations can be separated on the basis of size in Mg^{2+}-containing sucrose density gradients. A parallel centrifugation in EDTA will lead to dissociation of polysomes with a concomitant shift of the sedimentation behaviour of a translated mRNA, whereas untranslated mRNAs usually exhibit an unchanged sedimentation profile in EDTA. Based on the UV_{260} profile of a cytoplasmic extract, it is therefore possible to determine the effect of various treatments or physiological stimuli on global translation. However, the most powerful aspect of the sedimentation analysis is the ability to fractionate the gradient and isolate RNA from each fraction which is then subjected to Northern or slot-blot analysis (Fig. 5.9). In this way, it is feasible to examine the translational status of individual mRNAs, and the blots can be re-probed for other mRNAs of interest or internal standards.

Since the extraction of polysomes must not jeopardise the integrity of RNA-protein interactions, denaturants cannot be included in the extraction procedure, so it is necessary to add ribonuclease

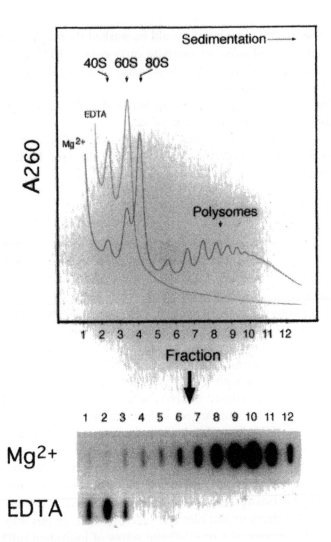

Fig. 5.9. Slot-blot analysis of GAPDH mRNA from a sucrose gradient fractionation. *Upper panel*: UV$_{260}$ profiles of a cytoplasmic lysate in a 20–47% (w/w) sucrose gradient containing either 5 mM Mg^{2+} (solid curve), or 10 mM EDTA (dotted curve). *Lower panel*: Autoradiograph from the slot-blot analysis of fractions from both gradients hybridised with a probe specific for glyceraldehyde 3-phosphate dehydrogenase. Notice the altered sedimentation behaviour of GAPDH mRNA as a result of EDTA-mediated release from ribosomes.

inhibitors such as RNasin or heparin to keep released ribonucleases at bay. Moreover, in most protocols the elongation inhibitor cycloheximide is usually added to cells before disruption and is also present in wash and lysis solutions to prevent ribosome 'run-off' during the preparation. A generally useful protocol for isolation and analysis of polysomal mRNAs from cell-lines is presented here, whereas extraction of polysomes from tissues such as placenta and pancreas presents a serious challenge that calls for specialised protocols.[19]

Materials
Phosphate-buffered saline
   10 mM NaH_2PO_4
   30 mM K_2HPO_4
   125 mM NaCl
   pH 7.4
Lysis solution
   20 mM Tris-HCl, pH 8.5
   1.5 mM $MgCl_2$
   140 mM KCl
   1 mM dithiothreitol
   0.1 mM cycloheximide
   0.5% Nonidet P-40
   1 unit/μl recombinant RNasin (Promega)
Low-density Mg^{2+} buffer
   20% w/w sucrose (Serva)
   20 mM Tris-HCl, pH 8.0
   5 mM $MgCl_2$
   140 mM KCl
High-density Mg^{2+} buffer
   47% w/w sucrose
   20 mM Tris-HCl, pH 8.0
   5 mM $MgCl_2$
   140 mM KCl

Low-density EDTA buffer
  20% w/w sucrose
  20 mM Tris-HCl, pH 8.0
  10 mM EDTA
  140 mM KCl
High-density EDTA buffer
  47% w/w sucrose
  20 mM Tris-HCl, pH 8.0
  10 mM EDTA
  140 mM KCl

Equipment
Gradient-mixer
Peristaltic pump
Ultracentrifuge
Swing-out bucket rotor (e.g. Beckman type SW41 Ti)

Procedure
1. Aspirate the medium from a 75 cm^2 culture flask containing 1–5 × 10^5 adherent cells per cm^2, and wash the cells with *phosphate-buffered saline* containing 10 μg/ml cycloheximide. Release the cells with 5 ml *phosphate-buffered saline* containing 2 g/l EDTA and 10 μg/ml cycloheximide at 37°C, pellet, and wash with ice-cold *phosphate-buffered saline* containing 10 μg/ml cycloheximide. Following centrifugation, cell pellets may be stored at −80°C.[a]
2. Prepare the 20–47% sucrose gradients in 14 × 89 mm polyallomer tubes by inserting a glass capillary in the bottom of the tube. The glass capillary is connected to the mixing chamber of a *gradient mixer* via a *peristaltic pump* and tubing. Add 6 ml of the appropriate (Mg^{2+} or EDTA) *low-density buffer* to the mixing chamber and 6 ml of the corresponding *high-density buffer* to the distal chamber. Open the valves and form the gradient. Insert the filled tubes in chilled *SW 41 buckets* and leave them on ice (see Section 4.2.3 for details).[b]

3. Lyse cell pellets from exponentially growing cells in 500 µl ice-cold *lysis solution* (use 1 ml if cells were confluent at harvest) by gently pipetting up and down a few times. Transfer to a microfuge tube and centrifuge at 10,000 g for 10 min in a refrigerated microfuge.

4. Apply 400 µl of the supernatant to the top of the sucrose gradient and centrifuge at 197,000 g (r_{av}) for 2 h and 15 min at 4°C (40,000 rpm in a *Beckman type SW 41 Ti rotor*).

5. Pump out the gradient carefully by inserting a glass capillary in the bottom of the polyallomer tube. The capillary is connected to a UV spectrophotometer and a fraction collector via a *peristaltic pump* and tubing.[c] Measure the absorbance at 260 nm and collect 1 ml fractions. Freeze fractions on dry ice.

6. Extract 500 µl of each fraction with phenol/chloroform (twice) and with chloroform.[d] Precipitate RNA by adding 1/10 vol 2.5 M sodium acetate (pH 6.0) and 2.5 vol ethanol, mix, and leave on dry ice for 20 min. Centrifuge at 10,000 g for 15 min in a refrigerated microfuge. Wash the pellet with 70% ethanol, re-centrifuge, and dry the pellet briefly under vacuum. Redissolve precipitated RNA in 50 µl double-distilled autoclaved H_2O and store at −80°C.

Notes

a. Cells in suspension culture are pelleted and washed with ice-cold *phosphate-buffered saline* containing 10 µg/ml cycloheximide.

b. The sucrose gradients are stable overnight.

c. The flow through the UV-recording flow-cell should be from top to bottom to avoid 'sucrose pulsation' in the flow-cell, and the base-line should be adjusted with the high-density solution before the gradient is pumped through (see Section 4.2.3, Chapter 4).

d. If fractions are to be analysed by slot-blotting on a Hybond-N membrane (Amersham), 1.5 ml 10 × SSPE, 6.25 M formaldehyde can be added directly to the phenol/chloroform-extracted aqueous phase in a 2 ml microfuge tube, followed by a 15 min treatment at 65°C.

Comments

The tedious work of extracting and analysing each of the twelve fractions is reduced considerably, if fractions containing mono-somes and subunits are combined separately from those containing polysomes. In doing so, potentially important information is sacr-ificed; namely, the ability to estimate shifts in the average number of ribosomes for a given size of reading frame, i.e. regulation of translation initiation.

Two parameters, besides the usual precautions when dealing with RNA, are important for obtaining good polysome profiles: (1) minimisation of ribosome 'run-off' during the initial preparation; and (2) inhibition of the released endogenous ribonucleases. The first problem is usually dealt with by a combination of rapid cooling and cycloheximide. However, cells should not be exposed to cy-cloheximide for longer than 5 min prior to harvest, since inhibition of translation elongation may trigger unknown feedback responses leading to artefacts of mRNA distribution. The second problem can be solved by recombinant RNasin that is preferable to earlier preparations of RNasin. A less expensive solution is the use of heparin, but in the initial stages of the preparation the level should not exceed 100 μg/ml which will lead to lysis of nuclei. Regardless of the inclusion of these additives, the polysome profiles from various cell-lines or tissues are different. The majority exhibit a fifty-fifty distribution between ribosomes in polysomes and the re-mainder in monosomes and subunits, but dispersed cells invariably contain more polysomes than corresponding growth-arrested cells.

There is a detailed review of the effect of ions and detergents on the release of subcellular mRNAs during lysis,[20] indicating that the inclusion of sodium deoxycholate is crucial for efficient release of membrane-bound polysomes. However, based on Northern analysis of the debris following lysis in 140 mM KCl, 1.5 mM MgCl$_2$

and 0.5% Nonidet P-40, the majority of cytoplasmic mRNA is released.

References

1. Moore, M.J., Query, C.C. and Sharp, P.A. (1993). Splicing of precursors to mRNAs by the spliceosome. In: The RNA World (Gesteland, R.F. and Atkins, J.F., eds.). Cold Spring Habor Laboratory Press, New York, pp. 303–358.
2. Grabowski, P.J., Padgett, R.A. and Sharp, P.A. (1984). Messenger RNA splicing *in vitro*: an excised intervening sequence and a potential intermediate. Cell *37*, 415–427.
3. Hernandez, N. and Keller, W. (1983). Splicing of *in vitro* synthesized messenger RNA precursors in HeLa cell extracts. Cell *35*, 89–99.
4. Dignam, J.D., Martin, P.L., Shastry, B.S. and Roeder, R.G. (1983). Eukaryotic gene transcription with purified components. Meth. Enzymol. *101*, 582–598.
5. Kramer, A., Frick, M. and Keller, W. (1987). Separation of multiple components of HeLa cell nuclear extracts required for pre-messenger RNA splicing. J. Biol. Chem. *262*, 17630–17640.
6. Berget, S.M. (1995). Exon recognition in vertebrate splicing. J. Biol. Chem. *270*, 2411–2414.
7. Konarska, M.M. and Sharp, P.A. (1986). Electrophoretic separation of complexes involved in the splicing of precursors to mRNAs. Cell *46*, 845–855.
8. Lamond, A.I. and Sproat, B.S. (1994). Isolation and characterization of ribonucleoprotein complexes. In: RNA Processing Vol. I (Higgens, S.J. and Hames, B.D., eds.). IRL Press, Oxford, pp. 103–140.
9. Ruskin, B. and Green, M.R. (1985). An RNA processing activity that debranches RNA lariats. Science *229*, 135–140.
10. Grabowski, P.J. and Sharp, P.A. (1986). Affinity chromatography of splicing complexes: U2, U5, and U4 + U6 small nuclear ribonucleoprotein particles in the spliceosome. Science *233*, 1294–1299.
11. Bindereif, A. and Green, M.R. (1987). An ordered pathway of snRNP binding during mammalian pre-mRNA splicing complex assembly. EMBO J. *6*, 2415–2424.
12. Kjems, J. and Sharp, P.A. (1993). The basic domain of Rev from human immunodeficiency virus type 1 specifically blocks the entry of U4/U6.U5

small nuclear ribonucleoprotein in spliceosome assembly. J. Virol. *67*, 4769–4776.

13. Moore, C.L. and Sharp, P.A. (1985). Accurate cleavage and polyadenylation of exogenous RNA substrate. Cell *41*, 845–855.

14. Bienroth, S., Keller, W. and Wahle, E. (1993). Assembly of a processive messenger RNA polyadenylation complex. EMBO J. *12*, 585–594.

15. Virtanen, A. and Sharp, P.A. (1988). Processing at immunoglobulin polyadenylation sites in lymphoid cell extracts. EMBO J. 7, 1421–1429.

16. Salles, F.J. and Strickland, S. (1995). Rapid and sensitive analysis of mRNA polyadenylation states by PCR. PCR Meth. Appl. *4*, 317–321.

17. Robertson, J.M. and Wintermeyer, W. (1981). Effect of translocation on topology and conformation of anticodon and D loops of tRNAPhe. J. Mol. Biol. *151*, 57–79.

18. Moldave, K. and Sadnik, I. (1979). Preparation of derived and native ribosomal subunits from rat liver. Meth. Enzymol. *59*, 402–410.

19. Kelly, S., Folman, R., Hochberg, A. and Ilan, J. (1980). Isolation of ribonuclease-free polysomes from human placenta. Biochim. Biophys. Acta *609*, 278–285.

20. Hesketh, J.E. and Pryme, I.F. (1991). Interaction between mRNA, ribosomes and the cytoskeleton. Biochem. J. *277*, 1–10.

Subject Index

Printed and bound by CPI Group (UK) Ltd, Croydon, CR0 4YY

03/10/2024

01040427-0018